Photonic Crystals - A Glimpse of the Current Research Trends

Edited by Pankaj Kumar Choudhury

Published in London, United Kingdom

IntechOpen

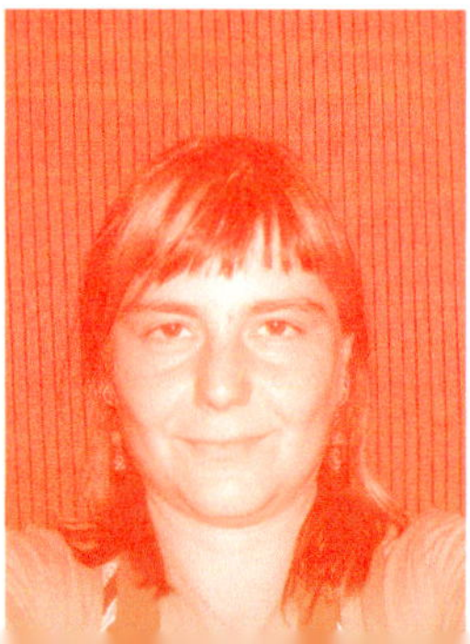

Supporting open minds since 2005

Photonic Crystals - A Glimpse of the Current Research Trends
http://dx.doi.org/10.5772/intechopen.76491
Edited by Pankaj Kumar Choudhury

Contributors
Ahmad Rifqi Md Zain, Yu-Cheng Hsiao, Binbin Weng, Tahere Hemati, Suobin Li, Arafa Aly, Pankaj Kumar Choudhury

First published in London, United Kingdom, 2019 by IntechOpen
IntechOpen is the global imprint of INTECHOPEN LIMITED, registered in England and Wales, registration number: 11086078, The Shard, 25th floor, 32 London Bridge Street
London, SE19SG - United Kingdom
Printed in Croatia

British Library Cataloguing-in-Publication Data
A catalogue record for this book is available from the British Library

Additional hard and PDF copies can be obtained from orders@intechopen.com

Photonic Crystals - A Glimpse of the Current Research Trends
Edited by Pankaj Kumar Choudhury
p. cm.
Print ISBN 978-1-83962-266-3
Online ISBN 978-1-83962-267-0
eBook (PDF) ISBN 978-1-83962-268-7

For EU product safety concerns:
IN TECH d.o.o., Prolaz Marije Krucifikse Kozulić 3, 51000 Rijeka, Croatia, info@intechopen.com or visit our website at intechopen.com.

Meet the editor

Pankaj Kumar Choudhury has held academic/research positions at Banaras Hindu University (Varanasi, India), Goa University (Goa, India), Laval University (Quebec, Canada), Gunma University (Kiryu, Japan), Multimedia University (Cyberjaya, Malaysia), and Telekom RandD (Malaysia). He is currently a professor at the Institute of Microengineering and Nanoelectronics, Universiti Kebangsaan Malaysia (Bangi, Malaysia). His research is focused on photonics. He has published more than 220 research papers, a number of chapters in 12 research-level books, and edited and coedited four books. He serves on the editorial board of *Optik—International Journal for Light and Electron Optics* (Elsevier, The Netherlands) and is the Editor-in-Chief of the *Journal of Electromagnetic Waves and Applications* (Taylor & Francis, UK). He is also senior member of the IEEE, OSA, and SPIE.

Contents

Preface

Photonic crystals (PhCs) are, in general, periodic structures, wherein the propagation of waves can be modulated with the variations in the refractive index properties of the medium. The bang-gap characteristics of these have been very attractive, and could be exploited to develop multitudes of integrated optic (IO) devices. The present volume is primarily aimed at the discussions of some of the features of PhCs of a few different types that can be used in certain areas of application.

Within the context, apart from the introductory chapter on the fundamentals of these specialized mediums, Aly and Mehaney present a comprehensive study of one-dimensional PhCs, taking into account their perfect forms as well as those with defects. They perform numerical simulations to investigate the angular effects of the incidence wave, putting the emphasis on the role of ambient temperature to determine the band-gap characteristics. The use of such mediums suggests prominence in temperature sensing. Furthermore, PhCs may also be embedded with engineered metallic structures to achieve certain band-gap characteristics, and Li and Dou discuss such microstructures to manipulate the allowed and forbidden regions so that wide complete band gaps may be achieved. Their results indicate potentials of the developed structures in low-frequency vibration reduction-related issues.

Certain PhC structures infiltrated with liquid crystals need special mention. Many IO devices can be conceptualized exploiting these because liquid crystals are functional materials, which alter their properties under the influence of external parameters, such as electromagnetic fields and temperature. Hsiao touches on the relevant features of liquid crystal-infiltrated PhCs to emphasize the structure as prudent in realizing varieties of tuneable optical devices.

In yet another discussion, Hemati and Weng elaborate the gas-sensing technologies in the mid-infrared region exploiting certain forms of PhC structures. This remains important because the mid-infrared gas-sensing systems can exhibit significantly high sensitivity. The authors review the advancements of research in this arena and emphasize the potentials of the technique in the currently growing Internet of Things technology.

Zain and De La Rue present finite difference time domain simulation results of engineered PhC-based waveguides, wherein the PhC cavities are embedded in narrow photonic wire silicon-on-insulator-type waveguides. The achieved high reflectivity and possibilities of active tuning capability leave the investigators to conceptualize multifunctional integrated circuits on a single chip.

The aforementioned cursory view of the themes encapsulated in *Photonic Crystals—A Glimpse of the Current Research Trends* highlights some of the present-day research in the PhC arena that remains of interest. Apart from these, many other directions of PhC-based research have been reported in the literature, and the editor hopes to take up these in a future volume. Finally, the editor is assured that the current volume incorporating chapters by a number of leading scientists will be significantly useful for the PhC research community.

Pankaj Kumar Choudhury
Professor,
Institute of Microengineering and Nanoelectronics (IMEN),
Universiti Kebangsaan Malaysia,
(The National University of Malaysia),
Bangi, Selangor, Malaysia

Chapter 1

Introductory Chapter: Photonic Crystals–Revisited

Pankaj Kumar Choudhury

1. Introduction

Mirrors are important optical devices as these are greatly useful in applications, such as imaging, solar energy collection, and filtering. Also, these are indispensable in lasers in which the cavities constitute one of the components of prominence. Metallic and dielectric are the two common types of mirrors highly used in optics. Metallic mirrors reflect electromagnetic waves over a broad range of frequencies. But, these are not suitable for operations in the infrared (IR) and optical frequency regimes due to high absorption of optical power, thereby causing loss. As such, the bandwidth of operation of metallic mirrors (or guides) remains limited. However, dielectric medium-coated metallic guides can be used in the IR regime. Nevertheless, these have been proved to be inefficient for operations in the optical regime owing to the absorption loss in metals. As such, the use of conventional dielectric waveguides remains a preferred option for the transmission of optical frequencies.

Within the context of confinement of light waves in a cavity, the invention of photonic crystals (PhCs) opens up many possibilities of controlling the propagation of light. This made PhCs as the objects of intensive theoretical and experimental research. Properties of optical fibers having high-index core region surrounded by silica or air fall into the class of PhC; various forms of these have been vastly discussed in the literature [1]. It is interesting to note that PhCs having high-index cores possess many features of conventional optical fibers. However, PhCs exhibit photonic band-gap (PBG) effect, which relates to the forbidden regions in dispersion characteristics and transmission spectra—the feature that is distinct from the properties of high-index core optical fibers.

As PhCs have been a research topic in the frontline for quite some time, emphasizing the avenues of such specialized microstructures, the present chapter aims at throwing a glimpse of a few different forms of guides falling in the class of PhCs. Some of the novel applications of PhC-based structures and the current research trends in this specialized area are also touched upon.

2. Periodic band-gap structures

PhCs can be classified to be one-, two-, or three-dimensional (1D, 2D, and 3D) periodic structures (depending on the kinds of periodic variation), comprised of materials with different refractive indices and having the periods comparable with the wavelengths of operation [2, 3]. Based on the configurations, there would be varieties of structures that can be regarded to be in the class of PhCs. For example, materials having periodic structures can exhibit the effect of PBG, provided the periodicity

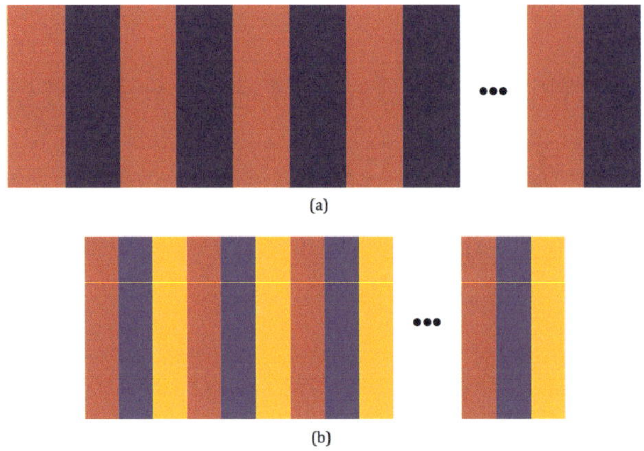

Figure 1.
Periodic stratified mediums with (a) two-layer and (b) three-layer periodicity.

remains on the scale of operating wavelength. Though the concept of periodic layered media had been put forward nearly over 40 years back [4, 5], varieties of new investigations in this front add many promising applications of such mediums. Mathematical formulations of symmetric periodic planar stratified mediums (**Figure 1a** and **b**) have been reported by the investigators [4]. The study deals with the evolution of explicit dispersion relations to determine the guided modes. In **Figure 1**, the periodicities in structure are of the forms of two (**Figure 1a**) and three layers (**Figure 1b**), respectively, which represent the variations in refractive index profiles. Apart from symmetric structures, the asymmetric dielectric film on a substrate is also quite a useful model among the more general types of optical waveguides.

The effects of PBG may be realized by the appropriate choice of periodic configurations, which involves dimensional features as well as the properties of constituent materials. This would result in forbidding the propagation of electromagnetic waves (through the structure) in certain frequency bands. It must be mentioned at this point that, recalling the Kronig-Penney model, the propagation of light waves in stratified periodic mediums can be compared with the situation of the movement of electrons in a periodic potential well [6, 7]. As such, if electrons can be diffracted by a periodic potential well, as evidenced by the theories of solid state physics, photons could equally be well-diffracted by a periodic modulation of the refractive index of medium. That is, such PBG mediums can be analyzed by exploiting the quantum theory of electrons in solids. This is because the basis for the guidance of light waves in dielectric mediums has a close analogy with the propagation of electrons in solid crystals—the fact that caused tremendous interests in PhCs and related research leading to the development of wide range of photonic structures for many novel applications [8]. In fact, multi-layered structures can reflect electromagnetic waves, if the frequency of operation lies within the gap. As such, stratified periodic structures have been proved to be prudent in exhibiting the property of spectral filtering.

3. Band-gap fibers

A PhC fiber (PCF) is a class of 2D periodic structure, wherein the periodic variation occurs in the plane perpendicular to the fiber axis and an invariant structure along it. In PCFs, the core section has the refractive index above the effective index of the surrounding medium. The guidance of light waves happens due to the total internal reflection (TIR). PCFs exhibit band-gap characteristics and present promising optical properties, such as lower and flat dispersion over a very large range of wavelength and reduced optical nonlinearities. Apart from these, PCFs demonstrate transparency in the far IR regime of electromagnetic spectrum [9–11].

In certain PhC configurations, the clad region may be a matrix of different materials with high and low refractive index values, thereby forming a new hybrid material that greatly enhances the core-clad index difference [12]. Within the context, index-guiding and hollow-core are the two different kinds of PCF; the former one consists of a doped-solid dielectric or pure silica core placed inside an air-clad guide, whereas the latter kind confines light waves through the PBG effect. The use of PBG kind of PCF greatly helps in reducing optical nonlinearity and propagation loss.

Among the others, PCFs are highly advantageous in fiber-based device applications. The invention of fiber-based lasers remains of special mention in this context. For example, continuous-wave fiber Brillouin lasers have been reported before utilizing highly nonlinear PCFs [13], wherein simple Fabry-Perot resonator plays the role of cavity. Apart from this, multi-wavelength Brillouin-erbium fiber lasers based on exploiting PCF with a linear cavity Fabry-Perot design have also been reported in the literature [14]. Many different schemes have been implemented to achieve fiber lasers having varieties of features [15].

Tunability of lasing systems using PCFs may be achieved in different ways. For example, one may use stimulated Brillouin scattering in the configurations [16, 17]. Within the context, the use of liquid crystals would also be greatly advantageous as these mediums exhibit the property of birefringence. Being liquid crystals as functional materials, one may recall the physical and/or chemical properties of liquid crystal, which can be altered by externally applied fields [18]. Apart from this, liquid crystals get affected due to the variations in temperature as well. As such, the thermal and electrical tuning of liquid crystals would alter the spectral characteristics—the feature that may be exploited in fabricating tunable PCFs. In fact, PCFs may be infiltrated with liquid crystals, in order to achieve tunable band-gap features.

4. Omniguiding fibers

Omniguiding fibers generally assume structures having the core surrounded by dielectric cylindrical Bragg mirrors comprised of alternating layers of high and low refractive index values, thereby forming a 1D PBG configuration, as shown in **Figure 2**. These are also called as Bragg fibers. However, several forms of omniguiding fibers have been reported in the literature. In certain kinds, the core section may be solid dielectric (e.g., silica or Ge-doped silica). In the case of hollow-core Bragg fibers, the core may be filled up with air or any other gaseous medium, as shown in **Figure 3**. In these guides, light waves remain confined to the core region due to Bragg reflections from the dielectric mirrors. This is because the mirrors reflect a narrow range of wavelength within the angular range. As such, complete photonic band-gap regime exists in phase space above the light cone of the surrounding mediums [19].

The design of omniguiding Bragg fibers requires adjustments of parametric values, such as the core thickness and refractive index of the alternating high- and low-index surrounding layered mediums. The number of layers also plays important roles in determining the allowed and forbidden wavelengths, i.e., the band-gap conditions. Omniguiding fibers may be designed as single-mode structure with no polarization degeneracy and without azimuthal dependence. The core size and number of concentric layers in these fibers govern the guided wavelengths, optical loss, and the effective single-mode operation [20]. As such,

Figure 2.
Cross-sectional view of a typical omniguiding Bragg fiber comprised of periodic multi-layered dielectric mirrors.

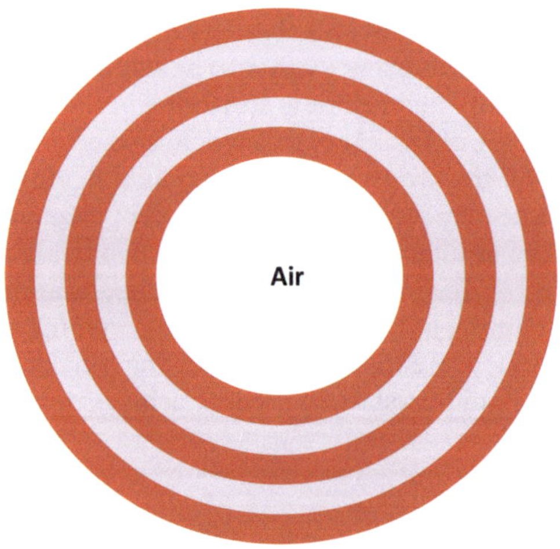

Figure 3.
Cross-sectional view of a hollow-core omniguiding fiber.

these may replace polarization maintaining fibers (PMFs)—the components that are used to eliminate the undesirable polarization dependent effects, such as polarization mode dispersion [21].

5. PhCs with defect

In the periodic configuration of PhC, certain defect units may be deliberately introduced that destroy the periodicity of medium. For example, if in the 1D PhC structure of **Figure 1a**, a defect layer (or unit, in general) is introduced, the configuration would assume the form, as shown in **Figure 4**. In such a case, the transmission characteristics of spectra will be drastically altered. In such situations, defect modes emerge inside the PBG, resulting into the presence of very narrow peaks with large transmissivity. As such, the transmission spectra of PhCs with defect can be controlled, provided the *introduced* defect is comprised of functional materials so that the electromagnetic behavior of these may be externally controlled.

Figure 4.
Periodic medium with a defect unit.

In general, the defect layer may be comprised of dielectric mediums or the mixture of dielectric and other kinds of mediums to yield a complex defect unit. In such situations, the transmission characteristics may be tailored in highly sophisticated ways, as described in ref. [22]. Apart from the planar structures, PCFs in certain forms may also have PhC cladding, wherein a central low-index structural defect would also be able to sustain the propagation of light waves.

6. Analytical approach for omniguiding fibers

Various attempts have been made to analyze the modes in omniguiding Bragg fibers. The most common approach remains as the use of transfer matrix theory that can be applied to any cylindrically symmetric fiber structure surrounded with periodically stacked Bragg cladding [23]. In this kind of formalism, the exact treatment of arbitrary number of inner dielectric layers is taken into account, and the structure of the outermost clad is approximated in the asymptotic limit. The exploitation of transfer matrix theory can yield the confined modes in Bragg fibers by minimizing the radiation loss in the radial direction. Apart from this technique, the asymptotic analysis and finite difference time domain (FDTD) method may also be used [3].

The method of asymptotic analysis of omniguiding fibers involves dividing the bulk of periodic multi-layered cylindrical dielectric mirrors into two groups, namely the inner and outer ones. The former one is in close proximity of the core section, whereas the latter group is assumed to be at relatively larger distance from the center of fiber. In the analyses, however, both the kinds of groups involve several dielectric mirrors. Further, the field in the inner group is represented by Bessel functions, whereas that in the outer group is treated asymptotically using the plane wave approximation. It has been found that the results obtained in this formalism match very well with those achieved by implementing the FDTD technique and/or the transfer matrix method [5].

In the analyses of omniguiding Bragg fibers, the propagation of Bloch waves is of extreme importance to determine the nature of propagation. Within the context, Bloch wave constant remains vital to evaluate as a complex value of it shows the forbidden bands of the periodic structure, whereas a real-valued Bloch constant indicates the propagation of waves. In the former case, however, the fields are evanescent.

Bragg fibers support modes that lie above the light line. These modes have the wave vector that corresponds to a frequency situated at the band-gap of multi-layered dielectric mirror. The imaginary part of wave vector indicates the radiative loss of modes that decreases exponentially with the increasing number of layers.

As stated before, the light wave propagation in Bragg fibers can be investigated in the analogy of electron flow in periodic lattice structures. This can be utilized in order to determine the working principle of omniguiding optical fibers. As such, the allowed and forbidden regions of these guides may be obtained by exploiting the quantum theory of electrons in solids. This has been justified that the use of simple Bloch formulation in omniguiding fibers exhibits continuous electric fields and power in dielectric boundaries [24, 25]. Furthermore, it has been found that, in the case of omniguiding fibers, the number of allowed and forbidden bands increases with the increase in the difference between the values of refractive index of different dielectric layers. Furthermore, the widths of the allowed band remain larger in the case of fibers having stacked layers of larger thickness values [26]. In the dispersion characteristics of omniguiding fibers as well, it has been found that the width of allowed range decrease with the increase in k/k_0, k and k_0 being the wave vector in the medium and that in the free-space, respectively. This has been demonstrated through obtaining thick curves [26] that represent the existence of allowed and forbidden bands of wavelengths, instead of simple lined curves shown by conventional optical fibers.

7. Recent research trends

With time, the research on PhC gained enough maturity, thereby yielding exciting results. R&D investigations report the possibilities of exploiting PhCs in many different applications. However, the aim of these remains pivoted to the tailoring of band-gap characteristics in the desired range of frequencies.

As stated in the preceding section, defects introduced in 1D PhC structures modify the propagation of waves, because such modes are governed by the PBG of medium [27, 28]. In this stream, the role of functional materials remains highly demanding as the form of PhC allows the possibility of tuning the spectral characteristics. The external effects, such as electromagnetic fields, temperature generating elastic, and/or shear waves, would be varied to achieve the desirable optical (or electromagnetic, in general) features of PhC. Therefore, hybrid PhCs would be useful in designing tunable optical filters, modulators, pulse compressors, and many others.

Apart from functional materials, metals may also be embedded in 1D PhCs to modify the confinement of modes, thereby altering the band-gap characteristics [29]. Interestingly, PhC structures embedded with metal-matrix arrangements could be used to reduce the low-frequency vibration and noise related issues.

Since PhCs can be exploited to control the light-matter interactions within micro/nano scales, these are advantageous for gas analyzing purpose [30–32]. In particular, the mid-infrared region of electromagnetic spectrum can be utilized for gas sensing applications, which would yield the development of such devices with high sensitivity [33, 34]. Indeed, the variations of optical spectrum and/or measuring the material properties are the techniques to determine the features of sensing.

PhC cavities can also be grown in nano-scaled photonic wire waveguides based on silicon-on-insulator (SOI). Such structures are capable to exhibit high reflectivity, which make them sophisticated candidates for mirroring in PhC structures [35–37]. These have been proved to be useful to realize active tuning—the feature that makes these suitable as basic building blocks for high-density photonic integrated circuits. Apart from this, the efficacy in designing filters and high-speed optical switches for networking applications have also been conceptualized.

The aforementioned features of PhCs describe only a few of the research ventures where these complex structures have been investigated. In reality, however, there are many other novel areas of research pivoted to exploiting varieties of new forms of PhCs to demonstrate fantastic electromagnetic features; all of those scopes remain beyond the coverage of this volume.

8. Summary

In analogy to the propagation of electron waves in periodic lattice structures, waves propagating in a structure that is periodically modulated with refractive

DOI: http://dx.doi.org/10.5772/intechopen.85246

index also exhibit photonic bands. Such *periodic* structures, comprised of high refractive index difference materials, yield photonic bands separated by gaps, thereby disallowing the propagation of waves. This triggers many novel approaches to manipulate the electromagnetic fields, thereby opening up varieties of possible technological applications.

Author details

Pankaj Kumar Choudhury
Institute of Microengineering and Nanoelectronics, Universiti Kebangsaan Malaysia, UKM, Bangi, Selangor, Malaysia

*Address all correspondence to: pankaj@ukm.edu.my

References

[1] Joannopoulos JD, Johnson SG, Winn JN, Meade RD. Photonic Crystals–Modelling the Flow of Light. 2nd ed. Princeton, New York: Princeton University Press; 2008

[2] Sakoda K. Optical Properties of Photonic Crystals. Berlin: Springer; 2005

[3] Passaro VMN. Advances in Photonic Crystals. UK: IntechOpen; 2013

[4] Yeh P, Yariv A, Hong CS. Electromagnetic propagation in periodic stratified media I–General theory. Journal of the Optical Society of America. 1977;**67**:423-438

[5] Yeh P, Yariv A, Marom E. Theory of Bragg fibre. Journal of the Optical Society of America. 1978;**68**:1196-1201

[6] Ojha SP, Choudhury PK, Khastgir P, Singh ON. Operating characteristics of an optical filter with a linearly periodic refractive index pattern in the filter material. Japanese Journal of Applied Physics. 1992;**31**:281-285

[7] Choudhury PK, Khastgir P, Ojha SP, Mahapatra DK, Singh ON. Design of an optical filter as a monochromatic selector from atomic emissions. Journal of the Optical Society of America A. 1992;**9**:1007-1010

[8] Shabanov VF, Vetrov SY, Shabanov AV. Optics of Real Photonic Crystals. Mesomorphic Defects and Inhomogeneities. Novosibirsk: Russian Academy of Sciences; 2005 in Russian

[9] Engeness TD, Ibanescu M, Johnson SG, Weisberg O, Skorobogatiy M, Jacobs S, et al. Dispersion tailoring and compensation by modal interactions in omniguide fibers. Optics Express. 2003;**11**:1175-1196

[10] Yablonovitch E. Photonic crystals. Journal of Modern Optics. 1994;**41**:173-194

[11] Russell P. Photonic crystal fibers. Science. 2003;**299**:358-362

[12] Rahman MM, Choudhury PK. On the investigation of field and power through photonic crystal fibers–A simulation approach. Optik–International Journal for Light and Electron Optics. 2011;**122**:963-969

[13] Lee JH, Yusoff Z, Belardi W, Ibsen M, Monro TM, Richardson DJ. Investigation of Brillouin effects in small-core holey optical fiber: Lasing and scattering. Optics Letters. 2002;**27**:927-929

[14] Nasir MNM, Yusoff Z, Al-mansoori MH, Rashid HAA, Choudhury PK. Photonic crystal fiber based multi-wavelength Brillouin-erbium laser. Proceedings of AOE. 2008;**2008**:SaG2e

[15] Desurvire E. Erbium-Doped Fiber Amplifiers. New York: Wiley; 1994

[16] Nasir MNM, Yusoff Z, Al-mansoori MH, Rashid HAA, Choudhury PK. Widely tunable multi-wavelength Brillouin-erbium fiber laser utilizing low SBS threshold photonic crystal fiber. Optics Express. 2009;7:12829-12834

[17] Nasir MNM, Yusoff Z, Al-mansoori MH, Rashid HAA, Choudhury PK. On the pre-amplified linear cavity multi-wavelength Brillioun-erbium fiber laser with low SBS threshold highly nonlinear photonic crystal fiber. Laser Physics. 2009;**19**:2027-2030

[18] Choudhury PK. New Developments in Liquid Crystals and Applications. New York: Nova; 2013

[19] Ibrahim A-BMA, Choudhury PK. Omniguiding Bragg fibers–Recent advances. In: Arkin WT, editor. Frontiers in Lasers and Electro-Optics Research. USA: Nova; 2006. pp. 63-88

[20] Argyros A. Guided modes and loss in Bragg fibres. Optics Express. 2002;**10**:1411-1417

[21] Bassett IM, Argyros A. Elimination of polarization degeneracy in round waveguides. Optics Express. 2002;**10**:1342-1346

[22] Wu J, Gao J. Defect mode in a one-dimensional photonic crystal with a dielectric-superconducting pair defect. Proceedings of SPIE. 2012;**8556**:85561Q

[23] Choudhury PK, Singh ON. Some multilayered and other unconventional lightguides. In: Singh ON, Lakhtakia A, editors. Electromagnetic Fields in Unconventional Structures and Materials. USA: Wiley; 2000. pp. 289-357

[24] Ibrahim A-BMA, Choudhury PK, Alias MS. On the analytical investigation of fields and power patterns in coaxial omniguiding Bragg fibers. Optik–International Journal for Light and Electron Optics. 2006;**117**:33-39

[25] Ibrahim A-BMA, Choudhury PK. Relative power distributions in omniguiding photonic band-gap fibers. Progress in Electromagnetics Research. 2007;**72**:269-278

[26] Ibrahim A-BMA, Choudhury PK, Alias MSB. Analytical design of photonic band-gap fibers and their dispersion characteristics. Optik–International Journal for Light and Electron Optics. 2005;**116**:169-174

[27] Painter O, Vuckovic J, Scherer A. Defect modes of a two-dimensional photonic crystal in an optically thin dielectric slab. Journal of the Optical Society of America B. 1999;**16**:275-285

[28] Dadoenkova NN, Zabolotin AE, Dadoenkova YS, Lyubchanskii IL, Lee YP, Rasing T. One-dimensional photonic crystals with the superconducting defects. In: Bananej A, editor. Photonic Crystals. UK: IntechOpen; 2015. pp. 1-25

[29] Li S, Chen T, Wang X, Xi Y. Lamb waves propagation in a novel metal-matrix phononic crystals plate. Modern Physics Letters B. 2016;**30**:1650338-1-1650338-13

[30] Snow P, Squire E, Russell PSJ, Canham L. Vapor sensing using the optical properties of porous silicon Bragg mirrors. Journal of Applied Physics. 1999;**86**:1781-1784

[31] Hodgkinson J, Tatam RP. Optical gas sensing–A review. Measurement Science and Technology. 2012;**24**:012004

[32] Zhang Y, Zhao Y, Lv R. A review for optical sensors based on photonic crystal cavities. Sensors and Actuators A: Physical. 2015;**233**:374-389

[33] Soref R. Mid-infrared photonics in silicon and germanium. Nature Photonics. 2010;**4**:495-497

[34] Lambrecht A, Hartwig S, Schweizer SL, Wehrspohn RB. Miniature infrared gas sensors using photonic crystals. Proceedings of SPIE. 2007;**6480**:64800D-1-64800D-10

[35] Notomi M, Shinya A, Yamada K, Takahashi J, Takahashi C, Yokohama I. Single-mode transmission within photonic bandgap of width-varied single-line-defect photonic crystal waveguides on SOI substrates. Electronics Letters. 2001;**37**:293-295

[36] Sakai A, Hara G, Baba T. Propagation characteristics of ultrahigh-delta optical waveguide on silicon-on-insulator substrate. Japanese Journal of Applied Physics. 2001;**40**:L383-L385

[37] Xia F, Sekaric L, Vlasov YA. Ultracompact optical buffers on a silicon chip. Nature Photonics. 2007;**1**:65-71

Chapter 2

Phonic Crystals and Thermal Effects

Arafa H. Aly and Ahmed Mehaney

Abstract

In this work, we demonstrate a comprehensive theoretical study of one-dimensional perfect and defect phononic crystals. In our study, we investigate the elastic and shear waves with the influences of thermal effects. The numerical calculations based on the transfer matrix method (TMM) and Bloch theory are presented, where the TMM is obtained by applying the continuity conditions between two consecutive sub-cells. Also, we show that by introducing a defect layer in the perfect periodic structures (defect phononic crystals), we obtain localization modes within the band structure. These localized modes can be implemented in many applications such as impedance matching, collimation, and focusing in acoustic imaging applications. Then, we investigate the influences of the incident angle and material types on the number and intensity of the localized modes in both cases of perfect/defect crystals. In addition, we have observed that the temperature has a great effect on the wave localization phenomena in phononic band gap structures. Such effects can change the thermal properties of the PnCs structure such as thermal conductivity, and it can also control the thermal emission, which is contributed by phonons in many engineering structures.

Keywords: thermal emission, dispersion relation, phononic band gap, localized modes

1. Introduction

Phononic crystals (PnCs) are new composite materials which can interact, manipulate, trap, prohibit, and transmit the propagation of mechanical waves. Recently, great efforts have been dedicated to study these novel materials to be used in many potential engineering applications. By reference to the meaning of idiomatic, the term "phononic" was derived in analogy to the term "phonon," which is considered as a quantization of the lattice vibrations. In the previous conscious of science, these vibrations are impossible to be controlled because the atoms in the solid cannot move independently of each other because they are connected by chemical bonds. By appearing PnCs, this assumption was changed and the mechanical wave can be filtered, transmitted, stopped, and localized within specific frequencies called phononic band gaps [1, 2].

Within the phononic band gaps, the mechanical waves of all types are greatly blocked. Actually, the formation of the phononic band gaps is back to the variation in the mechanical properties of the materials that build the PnC structure. Therefore, Bragg interference at the interface between each two materials can be obtained. As a result, the phononic band gaps and transmission bands are formed. Moreover, novel properties such as negative refraction and acoustic metamaterials are presented in PnC structures [3–7]. These novel properties of PnCs can be utilized in many industrial and engineering applications such as MEMS applications, filters, waveguides, clocking, multiplexers, and sensor applications [8–12].

As it is well-known, the physical origin of all crystal structures has the same idea of the design. Photonic crystals and semiconductors devices could not develop such types of the previous applications. From the scientific point of view, we should mention that the physical nature of PnCs is different from one of photonic crystals, as well as semiconductors. Generally, the various forms of waves are referred to as: electron waves as scalar waves and optical waves as vector waves, while elastic waves as tensor waves [13–19].

As a result, to design PnCs having a complete phononic band gap, the mechanical properties must be changed not only in one direction (i.e. x- direction) but also in all three directions of space (i.e., x, y, and z directions). Therefore, PnCs can be classified according to the periodicity into three types, i.e., the one-dimensional (1D), the two-dimensional (2D), and the three-dimensional (3D) PnCs [20–24]. PnCs can control the entire spectrum of phonons frequency from 1 Hz to THz range. As it is well-known, the waves that propagate through solid media are called elastic waves, while those that propagate through fluids are called acoustic waves. Also, PnCs can control the different types of mechanical waves such as elastic waves, acoustic waves, and surface waves. Therefore, unlike photonic crystals, not all 3D PnCs possess complete phononic band gaps. PnCs must possess band gaps for both elastic and acoustic waves at the same frequency region. Consequently, we have to fabricate PnC structures for both solid and fluid media [25–28]. Hence after, these different polarizations introduce more challenges to PnCs higher than other crystal, which in turn makes the theoretical manipulation of PnCs more attractive and perspective.

From the previous observations of PnCs structures, several methods were developed to calculate the phononic band gaps such as Plane-Wave-expansion Method (PWM), Bloch-Floquet Method (BFM), and finite different time-domain method (FDTD) [29–33]. In this chapter, we will depend on the Transfer Matrix Method (TMM) [34] for calculating the reflection coefficient and the transmission coefficient of the one-dimensional PnC structure. Such method considered very suitably for the 1D structure due to its recursive nature, since it allows the continuity of the waveform at the interface between each two layers [35, 36]. Also, by using the TMM, we can obtain the dispersion relations of the mechanical waves through the periodic PnC structures. PnCs and temperature are in mutual influences in several ways [37]. For example, at any temperature, mainly the huge contribution in the thermal conductivity of many materials is dominant by phonon contribution, which is a function of phonon mean free path and the Boltzmann distribution of phonons of any material. Also, the thermal conductivity is depending on the thickness of a material, where optical branches contribute with about 30% of the thermal conductivity of any material. By inhibiting the acoustic phonon population, the optical phonon relaxation is indirectly inhibited by up to 30% and hence limits their contribution to thermal conductivity. Furthermore, in silicon PnC, the thermal contribution of phonons has been reduced to less than 4% of the value for bulk silicon at room temperature [38–40].

Moreover, the defected PnCs structures have wonderful application in wave guiding and multiplexing. But the produced structures are different from the proposed ideal structures due to the errors and defaults in manufacturing. By removing some layers or materials from the ordered periodic structure in a PnC, we can create a point or a waveguide defects that are able to localize and bend signals. Such mechanism has been implemented similarly in photonic crystals to slow light. Consequently, the produced waveguides in PnCs structures can be used in focusing and collimation of acoustic waves in medical ultrasound applications, sensors, and MEMS applications [2, 41–44].

In the present work, we introduce the formation of phononic band gaps under the influences of many physical parameters. First, the general case of SH-wave propagation in arbitrary direction will be investigated by using TMM with calculation of the dispersion curves as well. Also, the influences of the incident angle on the band gaps are analyzed and discussed. In addition, this work focuses on correlating and comparing the results of SH-waves with in-plane waves propagating normally to the structure. For the in-plane waves, the reflection coefficients for S- and P-waves are plotted and compared with the dispersion relations curves. Furthermore, we are demonstrating the wave localization phenomenon in PnCs and the effects of the temperature on the band structure of PnCs and the localized modes for both in-plane and SH-waves. Also, the numerical results are presented and discussed to investigate the effect of the defect layer on the wave localization modes inside the structures. Finally, the effects of the thickness and type of the defect layer material on the band gap structure had been discussed.

2. Theoretical analysis and numerical models

2.1 Equation of wave propagation

Figure 1 shows the schematic diagram of the 1D PnC crystal structure. The proposed crystal structure has an infinite number of the periodically arranged unit cells. The unit cell may include two or more layers; here, we propose that it is made by only two materials A and B, respectively. The two materials are labeled by the subscript $j = 1, 2$. Also, the thickness of the unit cell (the lattice constant) is $a = a_1 + a_2$. The thickness, Lame' constant, shearing modulus, Poisson's ratios, mass density, and Young's modulus of the two layers are denoted by equation $a_j,\ \lambda_j, \mu_j, \nu_j, \rho_j, E_j \left[E_j = \lambda_j(1+\nu_j)(1-2\nu_j)/\nu_j\right.$, respectively.

The governing equation of anti-plane shear waves (SH-waves) polarized in the z-direction propagated in the xy-plane can be written in the following form [45–47].

$$\mu_j \nabla^2 \varphi_j\left(x_j, y_j, t\right) + \sigma_j^{tx} \frac{\partial^2 \varphi_j^2\left(x_j, y_j, t\right)}{\partial x^2} = \rho_j \ddot{\varphi}_j\left(x_j, y_j, t\right) \quad (j = 1, 2), \tag{1}$$

where T is the temperature variation, β_j is the thermal expansion coefficients, $\varphi_j\left(x_j, y_j, t\right)$ is the displacement components along the z-direction, t is the time,

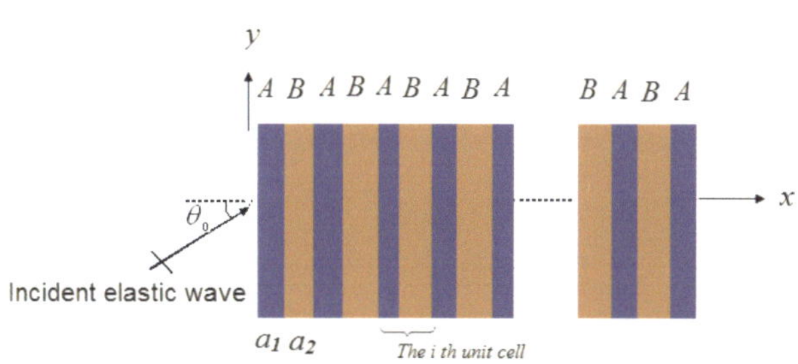

Figure 1.
A schematic diagram of a perfect 1D binary PnC structure.

$\sigma_j^{tx} = -E_j\beta_j T/(1-2\upsilon_j)$ is the thermal stress, and $\nabla^2 = \partial/\partial x_j^2 + \partial/\partial y_j^2$ is the Laplacian operator. The solution $\varphi_j\left(x_j, y_j, t\right)$ in the jth layer with time harmonic dependence can be expressed as [48],

$$\varphi_j\left(x_j, y_j, t\right) = \phi_j(x_j)\ \exp\left[iky_j \sin\theta_0 - i\omega t\right], \tag{2}$$

Where, c is the velocity component of the incident wave, $i^2 = -1$, $k = \omega/c$ is the wave number, θ_0 is the angle of the incident wave, ω is the angular frequency, and ϕ_j is the amplitude.

In many cases, it is more convenient to represent the thickness value by the following dimensionless coordinates:

$$\xi_j = {}^{x_j}/_{\overline{a}_1},\ \eta_j = {}^{y_j}/_{\overline{a}_1}\ (j = 1, 2), \tag{3}$$

where $\overline{a}_1$ is the average thickness value of the material A. $\overline{a}_1$ is equal to a_1 for the periodic structures. By inserting Eq. (3) in Eqs. (1) and (2), we can obtain the following wave equations:

$$\varphi_j\left(\xi_j, \eta_j, t\right) = \phi_j\left(\xi_j\right)\ \exp\left[i\alpha\eta_j \sin\theta_0 - i\omega t\right], \tag{4}$$

$$\left(1+\chi_j\right)\frac{d^2\phi_j}{d\xi_j^2} - \alpha^2\left(\sin^2\theta_0 - \frac{\alpha_j^2}{\alpha^2}\right)\phi_j = 0, \tag{5}$$

where c_j represents the wave velocity in each material, $\chi_j = \sigma_j^{tx}/\mu_j$ is the stress and shearing modulus ratio, $\alpha = k\overline{a}_1$ represents the SH-waves dimensionless wave number, $\alpha_j = k_j\overline{a}_1$ is the dimensionless wave number and $k_j = \omega/c_j$ is the wave vector of materials A and B, respectively. Eq. (5) has a general dimensionless solution, and it is given in the following form,

$$\phi_j\left(\xi_j\right) = A_j \exp\left(-i\alpha q_j \xi_j\right) + B_j \exp\left(+i\alpha q_j \xi_j\right), \quad (j = 1, 2), \tag{6}$$

where A_j and B_j are unknown coefficient to be determined, and q_j is a parameter and has the value $q_j = \frac{1}{\sqrt{1+\chi_j}}\sqrt{c^2/c_j^2 - \sin^2\theta_0}$. Therefore, the dimensionless solution $\varphi_j\left(\xi_j, \eta_j, t\right)$ with time harmonic dependence is:

$$\varphi_j\left(\xi_j, \eta_j, t\right) = \left(A_j \exp\left(-i\alpha q_j \xi_j\right) + B_j \exp\left(+i\alpha q_j \xi_j\right)\right)\left(\exp\left[i\alpha\eta_j \sin\theta_0 - i\omega t\right]\right). \tag{7}$$

2.2 Transfer matrix method

The dimensionless shear stress component is given as follows:

$$\tau_{xzj} = \mu_j \frac{\partial\varphi_j}{\overline{a}_1 \partial\xi_j} \quad (j = 1,\ 2). \tag{8}$$

Assuming that the PnC consists of n unit cells, the boundary conditions at the left and right sides of the two layers in the ith unit cell can be written in the following form [45],

$$\begin{gathered}\phi_{jL}^{(i)}=\phi_j^{(i)}(0),\quad \phi_{jR}^{(i)}=\phi_j^{(i)}\left(\zeta_j\right),\\ \tau_{xzjL}^{(i)}=\mu_j^{(i)}\frac{\partial\phi_j^{(i)}}{\bar{a}_1\partial\xi_j}(0),\quad \tau_{xzjR}^{(i)}=\mu_j^{(i)}\frac{\partial\phi_j^{(i)}}{\bar{a}_1\partial\xi_j}\left(\zeta_j\right)\quad (i=1,2,\dots..,n),\end{gathered}\tag{9}$$

where the subscripts L and R denote the left and right sides of the two layers, and $0\le\xi_j\le\zeta_j=a_j/\bar{a}_1\ \ (j=1,2)$ are the dimensionless thicknesses of materials A and B. Substituting Eqs. (7) and (8) into Eq. (9), the following matrix equation can be obtained as follows:

$$\nu_{jR}^{(i)}=T_j'\nu_{jL}^{(i)}\quad (j=1,2;i=1,2,\dots,n),\tag{10}$$

where $\nu_{jR}^{(i)}=\left\{\phi_{jR}^{(i)},\bar{a}_1\tau_{xzjR}^{(i)}\right\}^T$ and $\nu_{jL}^{(i)}=\left\{\phi_{jL}^{(i)},\bar{a}_1\tau_{xzjL}^{(i)}\right\}^T$ are the dimensionless state wave vectors at right and left sides of each unit cell and T_j' is the 2×2 transfer matrix of each unit cell. The four elements of T_j' can be written in the following forms,

$$\begin{aligned}T_j'(1,1)&=\frac{\exp\left(i\alpha q_j\zeta_j\right)+\exp\left(-i\alpha q_j\zeta_j\right)}{2},\quad T_j'(1,2)=\frac{\exp\left(i\alpha q_j\zeta_j\right)-\exp\left(-i\alpha q_j\zeta_j\right)}{2i\alpha q_j\mu^j j},\\ T_j'(2,1)&=\frac{i\alpha q_j\mu^j j\left[\exp\left(i\alpha q_j\zeta_j\right)-\exp\left(-i\alpha q_j\zeta_j\right)\right]}{2},\qquad T_j'(2,2)=T_j'(1,1).\end{aligned}\tag{11}$$

2.3 Characteristic of the dispersion relation

At the interface between the layers, the following condition is satisfied:

$$\nu_{1R}^{(i)}=\nu_{2L}^{(i)}.\tag{12}$$

Thus, the relationship between the right and left sides of the ith unit cell can be obtained from Eq. (10) as follows:

$$\nu_{2R}^{(i)}=T_i\nu_{1L}^{(i)}\quad (i=1,2,\dots,n),\tag{13}$$

where T_i is the accumulative transfer matrix of the ith unit cell and can be written in the following form:

$$T_i=T_2'T_1'.\tag{14}$$

At the interface between the right side of the unit cell and the left side of the ith unit cell, the following condition is satisfied:

$$\nu_{1L}^{(i)}=\nu_{2R}^{(i-1)}\quad (i=2,\dots\dots..,n).\tag{15}$$

By equating Eqs. (13) and (15), we can obtain the relationship between the state vectors of the $(i-1)$th unit cells and the ith unit cell in the following form:

$$\nu_{2R}^{(i)}=T_i\nu_{2R}^{(i-1)}\quad (i=2,\dots..,n),\tag{16}$$

Therefore, T_i represents the transfer matrix between each two successive unit cells.

By using Floquet and Bloch theories, we can obtain the displacement and stress fields between each two neighboring unit cells (at the interface) by the following relations [46–49]:

$$\phi_{2R}^{(i)} = \phi_{2R}^{(i-1)} \exp{(ika)}, \quad \tau_{xz2R}^{(i)} = \tau_{xz2R}^{(i-1)} \exp{(ika)}, \tag{17}$$

where k is the wave number representing the direction of the traveled wave across the structure. Combining the above two equations leads to the following matrix equation:

$$\nu_{2R}^{(i)} = \nu_{2R}^{(i-1)} \exp{(ika)} \quad (i = 2, \ldots.., n). \tag{18}$$

By equating Eqs. (16) and (18) which leads to the following eigenvalue problem:

$$\left|T_i - e^{ika} I\right| = 0. \tag{19}$$

Eq. (19) can be rewritten as follows:

$$T_i \nu_{2R}^{(i-1)} = \lambda\, \nu_{2R}^{(i-1)}, \tag{20}$$

where $\lambda = e^{ika}$ is a complex eigenvalue and $V_{2R}^{(i-1)}$ is a complex eigenvector.

2.4 Band structure formation

The wave number in Eqs. (17) and (18) is a complex number, so it can be a positive or negative number; in general, the wave number can be written in the following form [50]:

$$k = k_{\text{real}} - \mathrm{i}k_{\text{imaginary}}, \tag{21}$$

where k_{real} and $k_{\text{imaginary}}$ are the real and imaginary wave numbers, respectively. Therefore, we can deduce from this relation that the complex wave number has two forms, so it can inhibit the incident waves within the phononic band gaps. Consequently, we have two frequency ranges and they are organized as follows:

1. If $k = k_{\text{real}}$ and $k_{\text{real}} > 0$.

From Eq. (18),

$$\begin{aligned} v_{2R}^{(i)} &= v_{2R}^{(i-1)}\, \mathrm{e}^{ika}, \\ v_{2R}^{(i)} &= v_{2R}^{(i-1)}\, \mathrm{e}^{\mathrm{i}|k_{\text{real}}|a}. \end{aligned} \tag{22}$$

From Eq. (22), we can deduce that the displacement and stress at the successive unit cells (i)th and $(i-1)$th are differ only by a phase factor $\mathrm{e}^{\mathrm{i}|k_{real}|a}$. At this condition, the elastic waves are allowed to propagate freely through the PnC structure with the corresponding frequencies and wave number, forming the so-called transmission bands.

2. If $k = -\mathrm{i}k_{\text{imaginary}}$ and $k_{\text{imaginary}} < 0$

From Eq. (18),

$$\begin{aligned} v_{2R}^{(i)} &= v_{2R}^{(i-1)}\,\mathrm{e}^{ika}, \\ v_{2R}^{(i)} &= v_{2R}^{(i-1)}\,\mathrm{e}^{-|k_{\text{imaginary}}|a}. \end{aligned} \tag{23}$$

In contrast to the above case, we can deduce from Eq. (23) that the displacement and stress at successive unit cells (i)th and $(i-1)$th are the same and do not have a phase difference. Moreover, the incident wave has an exponential spatial attenuation of strength magnitude proportional to the value $|k_{\text{imaginary}}|$. Hence after, at this case, the waves are prohibited from propagation in the PnC structure, which in turn, results in formation of the so-called phononic band gaps or stop bands.

2.5 Temperature influences on PnCs

When a mechanical wave propagates through a PnC structure, resultant vibrations occur which may increase or decrease phonons motions. Therefore, it can heat or cool the PnC structure. Also, the thermal expansion of the constituents materials may change, which will affect the mechanical constants of the material as well. Since the vibrations occur very rapidly, there is no big chance to thermal energies to flow and the elastic constants measured by elastic waves propagation are changed adiabatically. The elastic constants are connected to the isothermal constants by the following relation [51–55],

$$\lambda^{\sigma} = \lambda^{\theta} + \frac{9\beta^2 B^2 \theta}{\rho C_\nu}, \quad \mu^{\sigma} = \mu^{\theta}, \tag{24}$$

where the superscripts σ and θ indicate adiabatic and isothermal constants, β is the thermal expansion coefficient, B is the bulk modulus$(B = \lambda + 2/3\mu)$, C_ν is the specific heat at constant volume, θ is the absolute temperature, and ρ is the mass density. From Eq. 24, we can deduce that λ^{σ} and λ^{θ} are not the same and the difference between them should be considered.

In addition to the above relations, if the temperature is increased, not only the wave velocities will increase but also the thickness of each layer will change by the following relation [53]:

$$\Delta a = \beta a_i \Delta t, \tag{25}$$

where Δa is the thickness difference of any layer, a_i is the original layer thickness, and Δt is the temperature difference. Hence after, these two variables will affect the longitudinal waves speed and the stress component in SH-waves equations. Since the P-wave velocity is $c_P = \sqrt{\frac{\lambda + 2\mu}{\rho}}$, which, in turn leads to the variation of the band structure and band gaps properties.

3. Numerical examples and discussions

3.1 SH-waves results

First, from a practical point of view, the number of layers in the 1D PnC structure should take a finite number. Therefore, we consider the unit cell of the PnC structure is made from two layers. The two layers are lead and epoxy materials and denoted by the symbols *A* and *B*, respectively. Second, dispersion relations are plotted for an infinite number of unit cells because it depends on Bloch theory that manipulates the propagation of waves through infinite periodic structures. Therefore, the dispersive behavior of the periodic materials and structures with an arbitrary chosen unit cell configuration is considered as an example. The constants of the materials used in the calculations can be found in Refs. [46, 49] and in **Table 1**. Here, we consider the velocity *c* of the incident wave is 800 m/s and the angle of the incidence is $\theta_0 = 20^\circ$. Also, the ambient temperature is proposed to be $T = 20^\circ C$. The dispersive properties of the inhomogeneous structures are determined by the properties ratios of the constituent materials, here we consider the two materials thicknesses ratio as 1:1.

Figure 2 presents the dispersion relation of SH-waves in the first Brillouin zone, and it is plotted between the nondimensional frequency $q = \omega a / c_B$ and the nondimensional wave number $\xi = k \times a$, where c_B is the wave velocity in the second material (epoxy). We considered the range of the nondimensional frequency is $0 \le q \le 9$. From this figure, we can conclude that, first, there are some

Materials	Mass density $\rho \times 10^3$ (kg/ m^3)	Lame' constant $\lambda \times 10^{10}$ (N/m^2)	Shearing modulus $\mu \times 10^{10}$ (N/m^2)	Young's modulus $E \times 10^{10}$ (N/m^2)	Poisson's ratios ν	Thermal expansion coefficient $\beta \times 10^{-6}$ (1/$^\circ$C)	Specific heat $C_\nu \times 10^3$ (J/kg. $^\circ$C)
Lead	11.4	3.3	0.54	1.536	0.43	29.5	0.128
Epoxy	1.180	0.443	0.159	0.435	0.368	22.5	1.182
Aluminum	2.699	6.1	2.5	6.752	0.355	23.9	0.9
Gold	19.32	15.0	2.85	8.114	0.42	14.2	0.13
Nylon	1.11	0.511	0.122	0.357	0.4	50	1.70

Table 1.
Material constants.

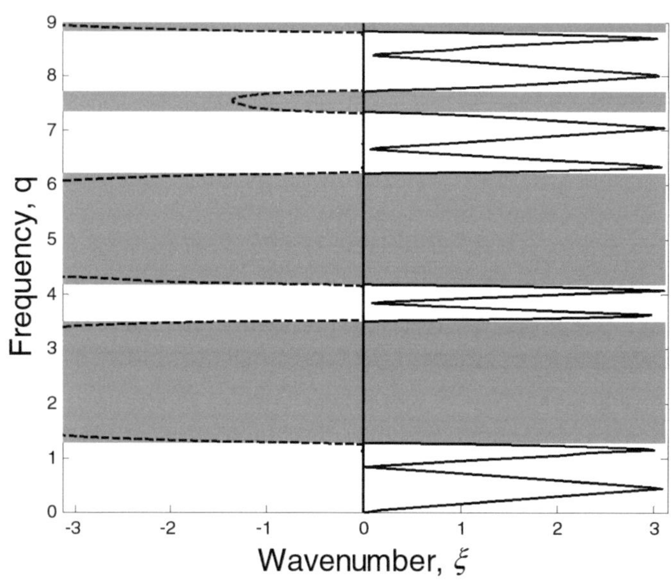

Figure 2.
The dispersion curve of SH-waves incident on the face of the 1D binary perfect PnC structure. Each unit cell consists of lead and epoxy materials (stop-bands shaded by the gray color).

frequency regions that are plotted with the white color and represent the real valued wave number and pass bands. Second, the other frequency regions corresponding to the imaginary wave number represent the phononic band gaps.

3.1.1 Influence of the angle of incidence on the phononic band gap

Figure 3(a) and **(b)** show the effects of the incident angle θ_0 on the band structures of the previous perfect structure. It can be seen that the wave propagation behavior of the perfect PnC changes obviously for different incident angles. For example, the stop bands for $\theta_0 = 15^\circ$ become a pass band for $\theta_0 = 45^\circ$. With increasing value of the incident angle, the same stop bands became wider in the considered frequency regions. However, in **Figure 3(c)**, we can note a wonderful phenomenon appeared when the angle θ_0 reached the value 85°. The stop band increased to be the entire band structure of the PnC, and no propagation bands were observed. The explanation of such phenomenon is deduced from the parameter $q_j = \frac{1}{\sqrt{1+\chi_j}}\sqrt{c^2/c_j^2 - \sin^2\theta_0}$; if c/c_j become smaller than $\sin\theta_0$, the displacement field will decay. Therefore, a total reflection to the elastic waves will be occurred, where all the incident wave energy is reflected back to the structure and no waves are allowed to propagate through the PnC structure.

3.1.2 Defective mode effect on the band structure of the PnC

As shown in **Figure 4**, the dispersion relation of the perfect PnC structure will differ than the defected ones. We consider a defect layer from aluminum (the defect layer thickness $a_d = a_A$ i.e., 1/2 of any unit cell thickness) was immersed after the second unit cell, and the material properties are mentioned in **Table 1**. From **Figure 5**, it was indicated that elastic waves can be trapped within the phononic band gap and the band structure changed significantly than the perfect ones. The width of the band gaps increased due to the increment of mismatch between the constituent materials, this back to the insertion of a second interface inside the PnC structure. Moreover, the defect layer acts as a trap inside the PnC structure, so a special wave frequency corresponding to that waveguide will propagate through the structure.

Additionally, in **Figure 6(a)** ($a_d = 2a_A$) and **Figure 6(b)** ($a_d = 4a_A$), we studied the effects of the defect layer thickness on the position and number of the localized modes inside the band structure of the PnC. It was shown that the number and width of the localized modes were increased by increasing the defect layer thickness. Therefore, with increasing the thickness of the defect layer inside the periodic PnC structure, the localized modes within the band gap are strongly confined within the defect layer.

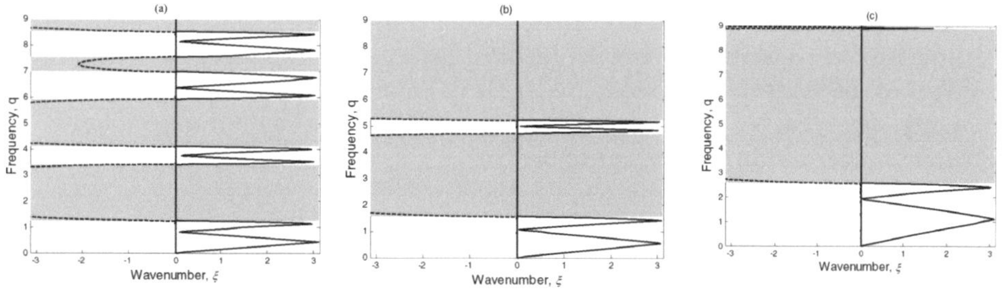

Figure 3.
The calculated dispersion curves of SH-waves in a 1D binary perfect PnC at the incident angles (a) $\theta_o^\circ = 30^\circ$, (b) $\theta_o^\circ = 50^\circ$, and (c) $\theta_o^\circ = 70^\circ$.

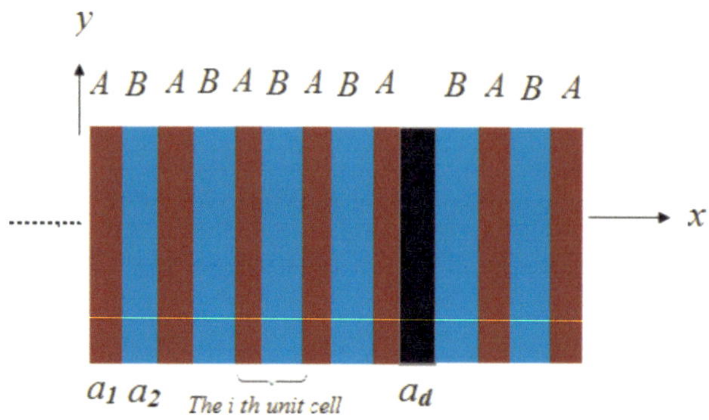

Figure 4.
A schematic diagram of a defect 1D binary PnC.

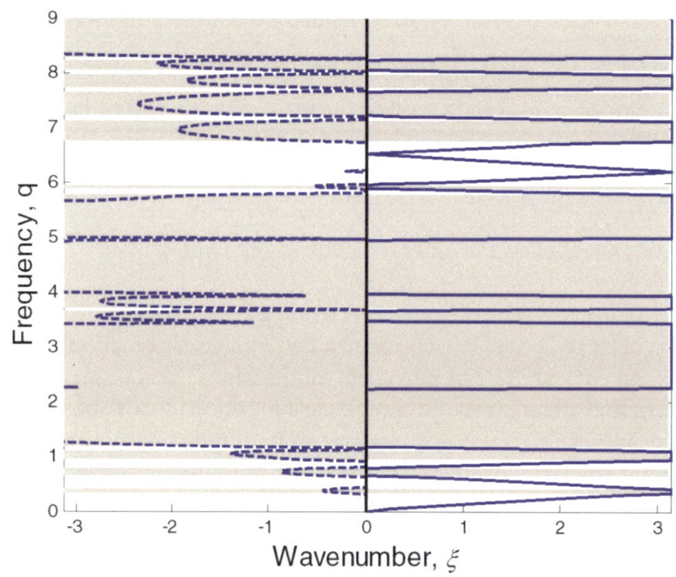

Figure 5.
The dispersion curve of SH-waves propagating in the defected PnC. A defect layer from aluminum with thickness $a_d = a_A$ was immersed between the two periodic unit cells (stop-bands shaded with the gray color).

Not only the thickness of the defect layer has an obvious effect on wave localization but also the type of the defect layer has a significant effect on the localized modes inside the band gaps. Although we introduced two different materials in **Figure 6(c)** and **(d)** with the same thickness, the number and width of the localized peaks had greatly changed. In **Figure 6(c)**, we used Au which has mechanical constants higher than the host materials, while in **Figure 5(d)**, we used nylon which has mechanical constants lower than the host materials. As a result, the width and number of the transmission bands are larger in **Figure 6(d)** than in **Figure 6(c)**. Actually, the nylon is a very soft material like a spring and can introduce more resonant modes inside the PnC structure.

3.1.3 Temperature effects on the band structure of PnC

Now we will investigate the effects of temperature elevation on the PnC structure and phononic band gaps. Two temperature degrees (T = 35 and 180$^{\circ}C$) were considered. **Figure 7** shows the response of the dispersion relations with

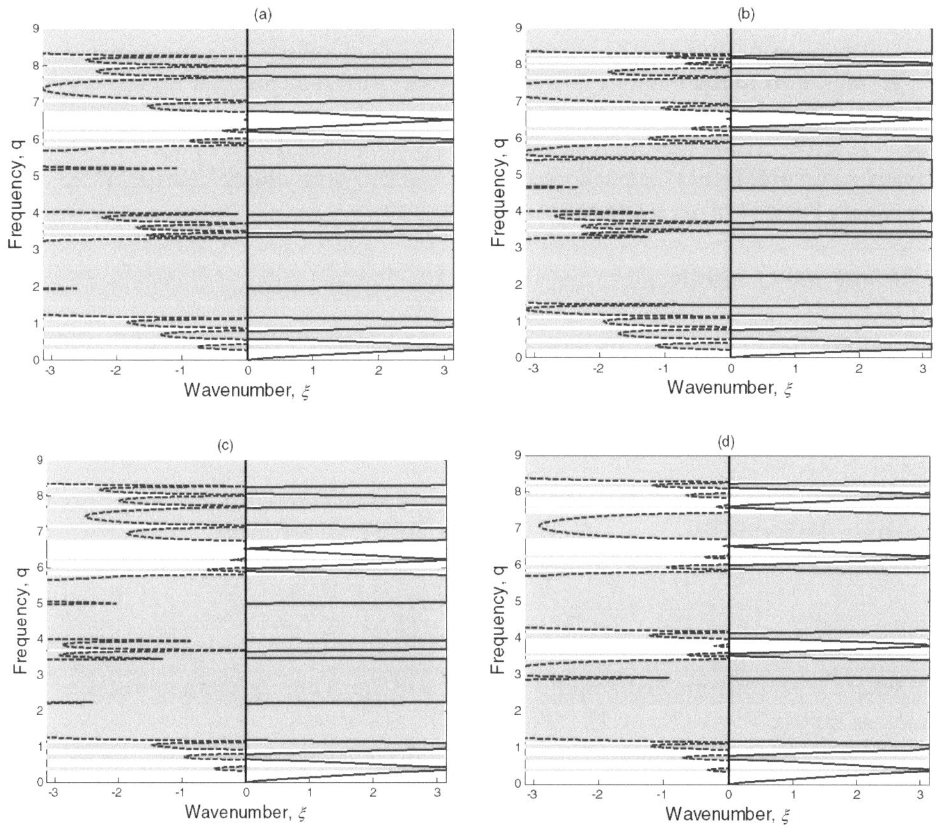

Figure 6.
The calculated dispersion curve of SH-waves propagating in a 1D binary defected PnC. (a) $a_d = 2a_A$ *(Al), (b)* $a_d = 4a_A$ *(Al), (c)* $a_d = a_A$ *(Au), and (d)* $a_d = a_A$ *(Nylon)* (stop-bands shaded with the gray color).

different temperatures at SH-waves propagation through PnCs. We can note in **Figure 7(a)** and **(b)** that, with increasing the temperature from $T = 35$ to $T = 180^{\circ}C$, respectively, the pass/stop band width remains constant due to the opposite increment in the thermal stress, which has a negative value and maintains materials dimensions constant. Therefore, there is not any variation in the thickness

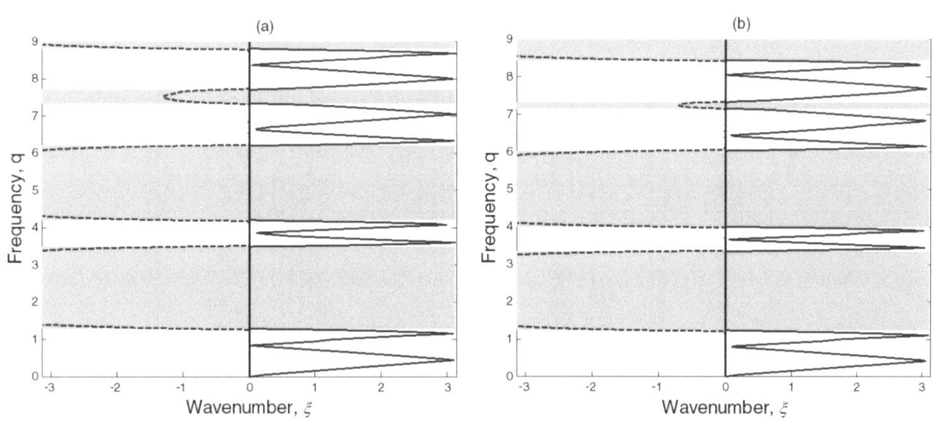

Figure 7.
The dispersion curve of SH-waves propagates in a PnC from lead and epoxy materials. Different temperatures are considered. (a) $T = 35^{\circ}C$ *and (b)* $T = 180^{\circ}C$ (stop-bands shaded in gray).

of any material, which in turn keeps the width of the band gaps constant. Finally, we can note only band edges changed slightly with increasing the temperatures. Such small shift in the band gaps is related to the different thermal expansion coefficients of the two materials. Based on this result, the effects of temperature on the PnC structure are greatly related to the type of waves that propagate through crystal structure. Hence, we will verify this result by studying the temperature effects on PnCs at plane wave propagation as well.

3.2 Plane waves results

As depicted in **Figure 1**, we can calculate the reflection coefficient of S- and P-waves in the x-direction through the PnC structure. The PnC structure is proposed to be bonded between two semi-infinite materials (nylon material) at the two ends. The subscripts "0" and "e" denote the left and the right of the PnC structure, respectively.

The reflection coefficient of the displacement field through a PnC structure is given by the form [39],

$$\frac{U_1}{U_0} = \frac{T_{12} + E_0 T_{11} - E_0 E_e T_{21} - E_e T_{22}}{E_0(T_{11} - E_e T_{21}) - (T_{12} - E_e T_{22})}, \tag{26}$$

where U_1 is the reflected amplitude and $T_{ij} = T(i,j)$ are the elements of the total transfer matrix $T = T_n T_{n-1} ... T_m ... T_1$.

Figure 8 shows the relation between the reflectance R versus $\omega a/_{2Pic_T}$ (ω is considered the angular frequency in this relation and $c_T = c_{SB}$) for P-wave (red dashed lines) and S-wave (black solid lines) [48]. From **Figure 8**, we can determine the range of frequencies for which the phononic band gaps can be occurred (reflectance of P- and S-waves is high $R \approx 1$). Such high reflectance within the different frequency ranges represents the frequency band gaps of P- and S-waves inside the PnC structure.

From **Figure 8**, it can be seen that the phononic band gaps described by the reflection coefficient are agreed with those described by the dispersion relations.

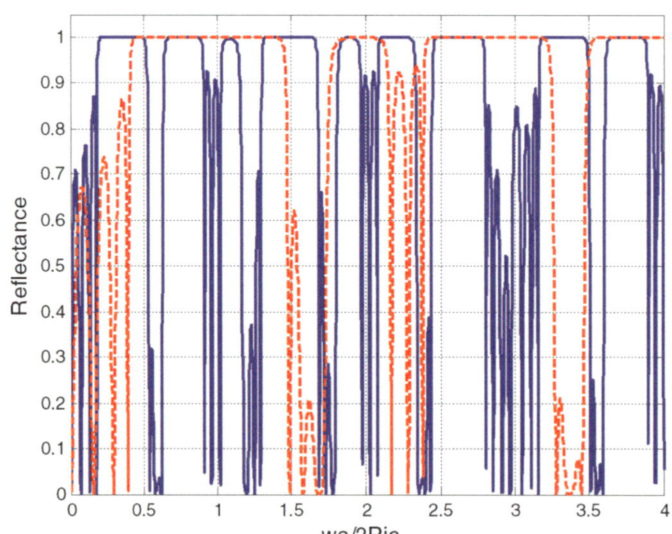

Figure 8.
The reflectance R *versus* $\omega a/2\pi c_T$ *of P-wave (red lines) and S-waves (blue lines) propagated normally through a 1D PnC structure consist of four unit cells. Each unit cell consists of lead and epoxy materials.*

The frequency regions at $R \approx 1$ of P- and S-waves represent the complete band gaps of the plane waves through the PnC structure. There is only a fine difference in the band gaps edges between **Figure 8** and those obtained by the dispersion relation. In **Figure 8**, the relation is plotted between the reflectance and $\omega a/_{2Pic_T}$, and $c_T = c_{SB}$ is chosen for both P- and S-waves reflectance. Therefore, we will use the reflection coefficient to describe the effects of the defect layer and temperature instead of dispersion relations in order to plot both P- and S-waves in the same graph and for the practical and industrial purpose as well.

3.2.1 Influences of the defect layer/temperature on the phononic band gaps "plane wave"

In this section, we will study the effects of the defect layer and temperatures on the band structure of PnCs at the propagation of plane waves and compare with those investigated for SH-waves.

3.2.1.1 Defect layer influences on band structure

First, we will use the same defected structure used in Section 3.1.2 with the same materials and conditions, only the angle of incidence will be maintained at $\theta_0 = 0^\circ$. **Figure 9** confirms the last results of the effects of the defect layer on the localization modes through the PnC at the plane wave propagation. In **Figure 9(b)**, a number of the localized waves was generated inside the phononic band gaps and was increased by increasing the defect layer thickness as well.

3.2.1.2 Temperature influences on band structure

In this section, the two temperatures $T = 50^\circ C$ and $190^\circ C$ were considered in order to illustrate the effects of temperature on the phonic band gaps. We noticed that the phononic band gaps were affected slightly by temperatures at plane wave propagation higher than SH-waves. Therefore, the propagation of elastic waves and localized modes can be affected by temperature elevation. From **Figure 10(a)**, we can note that the reflectance of the P-wave (Red lines) is moved toward the higher frequencies (i.e., band gap at $\omega a/2\pi c_T = 3.5$). Such displacement in the band gap edges is quite noticeable at $T = 190^\circ C$ in **Figure 10(b)**.

These temperature effects on the band gaps can be explained by two reasons. First, the P-wave velocity is increased according to Eq. (24) because the temperature has a direct effect on the elastic constants. Consequently, temperature makes a

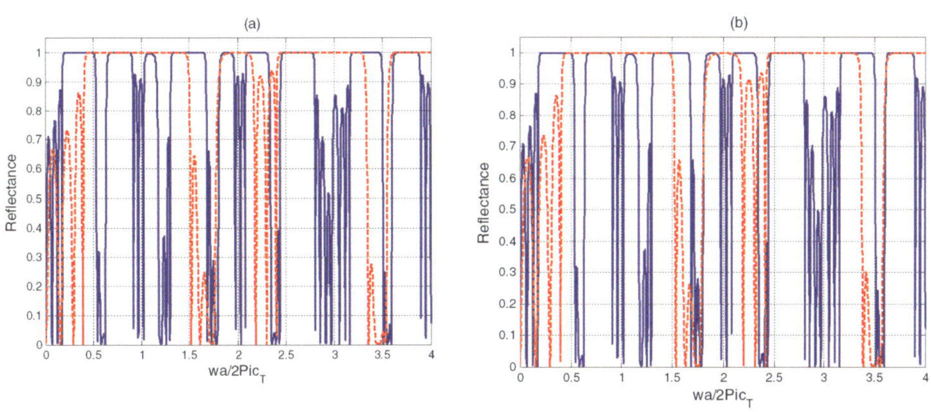

Figure 9.
(a) Aluminum defect layer with $a_d = a_A$, (b) Aluminum defect layer with $a_d = 4a_A$.

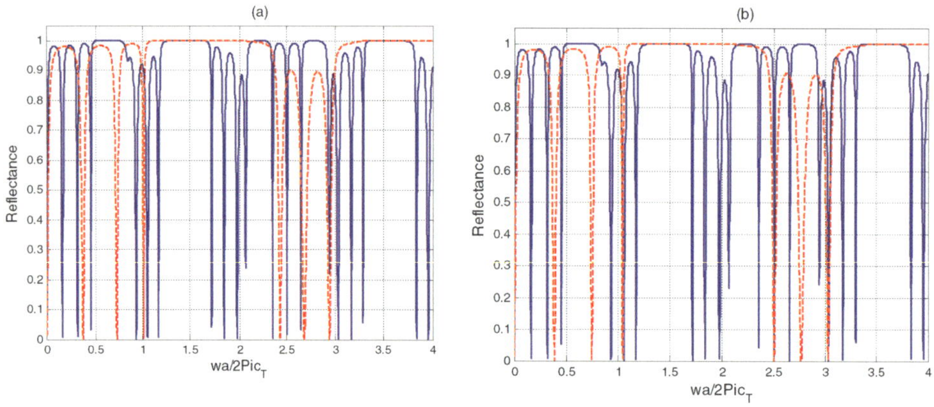

Figure 10.
The reflectance R versus $\omega a/2\pi c_T$ for P-wave (red lines) and S-waves (blue lines) at temperatures effects (a) T = 35°C and (b) T = 180°C.

notable change in the edges of the phononic band gaps. Second, here in the plane wave case, the thermal stress is absent, so the lattice constants will increase according to Eq. (25) and can make change in the width of the phononic band gaps as well.

4. Conclusions

In this chapter, we used the transfer matrix method and Bloch theory to compute and study the propagation of mechanical waves through different 1D PnC structures with thermal effects. We performed numerical simulations to calculate the reflection coefficient and dispersion curves for the 1D PnC structure. We have discussed and studied the effects of many parameters on the phononic band gaps. The effects of temperature and defect mode on the properties of the PnC and on the localized modes were studied for both SH-waves and in-plane waves. The output results can be stated as follows:

1. The defect layer has a different and unique effect on the number of localized modes inside the phononic band gaps for both plane and SH-waves. The effects of the incident angle, thickness ratio on the propagation, and localization of elastic waves were studied. The number of localized modes was increased with increasing the defect layer thickness. Also, the material type of the defect layer has an obvious effect on the width and number of resonant defect modes, where the localized modes were increased by using a material with low elastic constant. Inserting a defect layer inside the perfect structure facilitates the process of localized modes generation inside the PnC structure. Also, it can be utilized in many potential applications such as acoustic filter and waveguides and can be applied to produce new types of laser called phonon laser with low thermal effects especially in the hypersonic PnCs regime.

2. Results revealed that temperature increment has a significant effect on the phononic band gaps and on the number of the localized modes inside the PnC structure, especially for plane wave, where the P-waves velocity and lattice constants are directly related to temperature increment. The contribution of phonons in the thermal conductivity of a solid can be limited by the existence of phononic band gaps, which could be very useful for the thermoelectric

devices that convert thermal energy into electricity. Hence after, the performance of many devices such as Peltier thermoelectric coolers, thermocouples, sensors, and thermoelectric energy generators can be enhanced. Moreover, the thermal conductivity of many structures could experience great changes based on the temperature gradients of PnCs. When temperature is increase in PnCs, the phononic band gap experiences a significant change.

Conflict of interest

The authors declare that there is no conflict of interest.

Author details

Arafa H. Aly* and Ahmed Mehaney
Physics Department, Faculty of Sciences, Beni-Suef University, Egypt

*Address all correspondence to: arafaaly@aucegypt.edu

References

[1] Gorishyy T, Maldovan M, Ullal C, Thomas E. Sound ideas. Physics world. 2005;24-29

[2] Armenise MN, Campanella CE, Ciminelli C, dell'Olio F, passro VMN. Photonic and phononic band gap structures: modeling and applications. Physics Procedia. 2010;**3**:357-364

[3] Torrent D, Sanchez-Dehesa J. Radial wave crystals: Radially periodic structures from anisotropic metamaterials for engineering acoustic or electromagnetic waves. Physical Review Letters. 2009;**103**:064301

[4] Fok L, Zhang X. Negative acoustic index metamaterial. Physical Review B. 2011;**83**:214304

[5] Cai C, Wang Z, Chu Y, Liu G, Xu Z. The phononic band gaps of Bragg scattering and locally resonant pentamode Metamaterials. Journal of Physics D: Applied Physics. 2017;**50**:41

[6] Warmuth F, Wormser M, Körner C. Single phase 3D phononic band gap material. Scientific Reports. 2017;**7**:3843

[7] Jia Z, Chen Y, Yang H, Wang L. Designing Phononic Crystals with Wide and Robust Band Gaps. Physical Review Applied. 2018;**9**:044021

[8] Aly AH, Nagaty A, Khalifa Z. Propagation of acoustic waves in 2D periodic and quasiperiodic phononic crystals. International Journal of Modern Physics B. 2017;**31**

[9] Christensen J, Fernandez-Dominguez AI, de Leon-Perez F, Martin-Moreno L, Garcia-Vidal FJ. Collimation of sound assisted by acoustic surface waves. Nature Physics. 2007;**3**:851-852

[10] Aly AH, Mehaney A, Hanafey HS. Phononic band gaps in one dimensional mass spring system. PIERS Proceeding. 2012;**1047**:27-30

[11] Aly AH, Nagaty A, Khalifa Z, Mehaney A. The significance of temperature dependence on the piezoelectric energy harvesting by using a phononic crystal. Journal of Applied Physics. 2018;**123**:185102

[12] Aly AH, Mohamed D, Elsayed HA, Mehaney A. Fano resonance by means of the one-dimensional superconductor photonic crystals. Journal of Superconductivity and Novel Magnetism. 2018;**31**:3827-3833

[13] Aly AH, Mehaney A. Phononic crystals with one-dimensional defect as sensor materials. Indian Journal of Physics. 2017;**91**:1021

[14] El-Kady I, Olsson RH III, Fleming JG. Phononic band gap crystals for radio frequency communications. Applied Physics Letters. 2008;**92**:233504

[15] Mehaney A, Eissa MF, Aly AH. Detection and discrimination between alpha particles and protons based on phononic crystals materials. Surface Review and Letters. 2018. DOI: 10.1142/S0218625X18502190

[16] Aly AH, Nagaty A, Mehaney A. One-dimensional phononic crystals that incorporate a defective piezoelectric/piezomagnetic as a new sensor. The European Physical Journal B. 2018;**91**:211. DOI: 10.1140/epjb/e2018-90347-6

[17] Mohammadi S, Eftekhar AA, Hunt WD, Adibi A. High-Q micromechanical resonators in a two-dimensional phononic crystal slab. Applied Physics Letters. 2009;**94**:051906

[18] Nagaty A, Mehaney A, Aly AH. Influence of temperature on the

properties of one-dimensional piezoelectric phononic crystals. Chinese Physics B. 2018;**27**:094301

[19] Nagaty A, Mehaney A, Aly AH. Acoustic wave sensor based on piezomagnetic phononic crystal. Journal of Superconductivity and Novel Magnetism. 2018;**31**:4173-4177

[20] Chen JJ, Han X. The propagation of Lamb waves in one-dimensional phononic crystal plates bordered with symmetric uniform layers. Physics Letters A. 2010;**374**:3243-3246

[21] Chen A, Wang Y, Yu G, Guo Y, Wang Z. Elastic wave localization in two-dimensional phononic crystals with one-dimensional quasi-periodicity and random disorder. Acta Mechanica Solida Sinica. 2008;**21**:517-528

[22] Torrent D, Sanchez-Dehesa J. Acoustic resonances in two-dimensional radial sonic crystal shells. New Journal of Physics. 2010;**12**:073034

[23] Liu Z, Zhang X, Mao Y, Zhu YY, Yang Z, Chan CT, et al. Locally resonant sonic materials. Science. 2000;**289**: 1734-1736

[24] Kushwaha MS, Halevi P, Martinez G, Dobrzynski L, Djafari-Rouhani B. Theory of acoustic band structure of periodic elastic composities. Physical Review B. 1994;**49**:2313-2322

[25] Tan EL. Generalized eigenproblem of hybrid matrix for Floquet wave propagation in one-dimensional phononic crystals with solids and fluids. Ultrasonics. 2010;**50**:91-98

[26] Pennec Y, Djafari-Rouhani B, Vasseur JO, Khelif A, Deymier PA. Tunable filtering and demultiplexing in phononic crystals with hollow cylinders. Physical Review E. 2004;**69**: 046608

[27] Lucklum R, Li J, Zubtsov M. 1D and 2D phononic crystal sensors. Procedia Engineering. 2010;**5**:436-439

[28] Zhao D, Wang W, Liu Z, Shi J, Wen W. Peculiar transmission property of acoustic waves in a one-dimensional layered phononic crystal. Physica B. 2007;**390**:159-166

[29] Zhang X, Liu YY, Wu FG, Liu ZY. Large two-dimensional band gaps in three-component phononic crystals. Physics Letters A. 2003;**317**:144-149

[30] Ciminelli C, Peluso F, Armenise MN. Modelling and design of two-dimensional guided-wave photonic band-gap devices. IEEE.Journal of Light wave Technology. 2005;**23**:886-901

[31] Taflove A. Computational Electrodynamics: The Finite-Difference Time –Domain method. Norwood, MA: Artech House Inc; 1995

[32] Dwoyer DL, Hussaini MY, Voigt RG. Finite Element-Theory and Application. New York: Spriger-Verlag; 1986

[33] Zhao HG, Liu YZ, Wen JH, Yu DL, Wen XS. Tri-component phononic crystals for underwater anechoic coatings. Physics Letters A. 2007;**367**: 224-232

[34] Thomson WT. Transmission of elastic waves through a stratified solid medium. Journal of Applied Physics. 1950;**21**:89-93

[35] Liu L, Hussein MI. Wave motion in periodic flexural beams and characterization of the transition between Bragg scattering and local resonance. Journal of Applied Mechanics. 2012;**79**:011003

[36] Hussein MI, Hamza K, Hulbert GM, Scott RA, Saitou K. Multiobjective evolutionary optimization of periodic layered materials for desired wave

dispersion characteristics. Structural Multidisciplinary Optimization. 2006; **31**:60-75

[37] Aly AH, Mehaney A. Modulation of the band gaps of phononic crystals with thermal effects. International Journal of Thermophysics. 2015;**36**:2967

[38] Ciampa F, Mankar A, Marini A. Phononic crystal waveguide transducers for nonlinear elastic wave sensing. Scientific Reports. 2017;**7**:14712

[39] Wu LY, Yang WP, Chen LW. The thermal effects on the negative refraction of sonic crystals. Physics Letters A. 2008;**372**:2701-2705

[40] Hopkins PE, Phinney LM, Rakich PT, Olsson III RH, EL-Kady I. Phonon considerations in the reduction of thermal conductivity in phononic crystals. In: MeTA'10 2nd International Conference on Metamaterials, Photonic Crystals and Plasmonics. 2010. pp. 308-316

[41] Aly AH, Mehaney A, El-Naggar SA. Evolution of phononic band gaps in one-dimensional phononic crystals that incorporate high-Tc superconductor and magnetostrictive materials. Journal of Superconductivity and Novel Magnetism. 2017;**30**:2711-2716

[42] Chen AL, Wang Y-S. Study on band gaps of elastic waves propagating in one-dimensional disordered phononic crystals. Physica B. 2007;**392**:369-378

[43] Aly AH, Mehaney A. Low band gap frequencies and multiplexing properties in 1D and 2D mass spring structures. Chinese Physics B. 2016;**25**:114301

[44] Aly AH, Mehaney A, Eissa MF. Ionizing particle detection based on phononic crystals. Journal of Applied Physics. 2015;**118**:064502

[45] Rose JL. Ultrasonic Waves in Solid Media. London: Cambridge University Press; 1999

[46] Wang YZ, Li FM, Kishimoto K, Wang YH, Huang WH. Wave localization in randomly disordered layered three-component phononic crystals with thermal effects. Archive of Applied Mechanics. 2010;**80**:629-640

[47] Shang FL, Wang ZK, Li ZH. An exact analysis of thermal buckling of piezoelectric laminated plates. Acta Mechanica Solida Sinica. 1997;**10**:95-107

[48] Li FM, Wang YS. Study on wave localization in disordered periodic layered piezoelectric composite structures. International Journal of Solids Structures. 2005;**42**:6457-6474

[49] Hussein MI, Hulbert GM, Scott RA. Dispersive elastodynamics of 1D banded materials and structures: analysis. Journal of Sound and Vibration. 2006; **289**:779-806

[50] Aly AH, Mehaney A. Enhancement of phononic band gaps in ternary/binary structure. Physica B: Condensed Matter. 2012;**407**(21):4262-4268

[51] Gray DE, Fischel DN, Crawford HB, Douglas DA, Eisler WC. American Institute of Physics Handbook. 3rd ed. USA: Colonial Press, McGraw-Hill Book Company; 1972

[52] Mason WP. Piezoelectric Crystals and Their Application to Ultrasonics. Princeton, N. J: D. Van Nostrand Company, Inc; 1950. pp. 480-481

[53] Serway RA, Jewett JW. Physics for Scientists and Engineers. 6th ed. USA: Thomson Brooks/Cole; 2004

[54] Aly AH, Mehaney A, Rahman EA. Study of physical parameters on the properties of phononic band gaps. International Journal of Modern Physics B. 2013;**27**:1350047

[55] Mehaney A. Phononic crystal as a neutron detector. Ultrasonics. 2019;**93**: 37-42

Chapter 3

Metal-Matrix Embedded Phononic Crystals

Suobin Li, Yihua Dou and Linka Niu

Abstract

Metal-matrix embedded phononic crystals (MMEPCs) can be applied for noise and vibration reduction. Metal-matrix embedded phononic crystals (MMEPCs) consisting of double-sided stubs (single "hard" stubs/composite stubs) were introduced. The introduced MMEPCs are deposited on a two-dimensional locally resonant phononic crystal plate that consists of an array of rubber fillers embedded in a steel plate. The lower frequency complete bandgap will be produced in the MMEPCs with composite stubs by decoupling the spring-mass system of the resonator by means of the rubber filler. Then, the out-of-plane bandgap and the in-plane bandgap can be adjusted into the same lowest frequency range by the composite stubs. The broad complete bandgap will be produced in the metal-matrix embedded phononic crystals with single "hard" stubs by producing new kinds of resonance modes (in-plane and out-of-plane analogous-rigid modes) by introducing the single "hard" stubs, and then the out-of-plane bandgap and the in-plane bandgap can be broadened into the same frequency range by the single "hard" stubs. The proposed MMEPCs can be used for noise and vibration reduction.

Keywords: metal-matrix embedded phononic crystals, lower frequency complete bandgap, broad complete bandgap, forming mechanisms of the bandgap, noise, vibration reduction

1. Introduction

In the last two decades, the propagation of elastic waves in periodic composite materials, known as phononic crystals, has attracted increased attention due to their unique physical properties. These properties include phononic bandgaps which define a frequency range where elastic wave propagation is forbidden [1]. The existence of phononic bandgaps enables a variety of potential applications, such as noise and vibration insulation. Two mechanisms cause the formation of bandgaps: Bragg scattering [1] and local resonance [2]. In Bragg scattering, the associated wavelength with the phononic bandgap is of the same order as the periodicity of the structure. This means a huge lattice constant is needed to obtain phononic bandgaps for the low frequency range, which limits applications. To achieve local resonance, the associated wavelength needs to be two orders of magnitude smaller than the Bragg bandgap. A locally resonant bandgap is related to the resonance frequency associated with scattering units and depends less on the periodicity and symmetry of the structure. Therefore, it overcomes the limitation of Bragg bandgaps and permits bandgaps suitable for low frequencies.

The existence of low-frequency locally resonant bandgaps (LRBGs) gives rise to the application of locally resonant phononic crystals (LRPCs) in the reduction of low-frequency vibration and noise [3]. However, it has been a challenging task due to the long wavelength and weak attenuation of the low-frequency waves. In the past years, many LRPCs were proposed to obtain bandgaps in low frequency range or broaden bandgaps in the low frequency range [4–19]. More recently, the propagation of Lamb waves (elastic waves) in phononic crystal plates, which is based on a local resonance mechanism, has attracted attention due to their potential applications in filters, resonators, waveguides, and for vibration insulation. In general, phononic-crystal plates can be classified into two types according to their structural features: flat plates and stubbed plates. The flat phononic-crystal plate consists of a periodical array of holes or periodic inclusions of foreign material in a homogeneous plate [5]. Many previous studies focused on this type of phononic-crystal plates to investigate the bandgap properties. It is not easy to obtain lower frequency phononic bandgaps using a flat phononic-crystal plate. Stubbed phononic-crystal plates can be classified into two types: the single-sided stubbed phononic-crystal plate and the double-sided stubbed phononic-crystal plate. The single-sided stubbed phononic-crystal plate [3] consists of a square array of stubs on one side of a homogeneous plate [6]. Subsequently, many researchers investigated the effects of material properties and geometric parameters on the BG of the single-sided stubbed phononic-crystal plate and found that the mass and the geometric parameter effect can help tune the bandgap into low frequencies [8]. Later, a novel single-sided stubbed phononic-crystal plate was proposed. It consists of a square array of stubs on one side of a two-dimensional binary locally resonant phononic plate. It can increase of the relative bandwidth over the classical single-sided stubbed plate [12]. The double-sided stubbed phononic-crystal plate consists of a square array of stubs on both sides of a homogeneous plate. It can reduce the bandgap compared to a single-sided stubbed phononic-crystal plate [9]. Recently, a novel double-sided stubbed phononic-crystal plate was proposed. Unlike the classical double-sided stubbed phononic-crystal plate, it can shift the bandgap into the lower frequency range [13].

However, because there are three different modes for a plate, and they can only be coupled separately to special resonating modes of the resonators, the gaps for in-plane and out-of-plane plate modes can hardly be overlapped with each other in lower frequencies due to the coupling of the spring-mass system of the resonator, and thus the lower complete bandgaps (below 100 Hz) are difficult to obtain. As a result, the BGs of these disused PC plates are usually located in the frequency range above 300 Hz. However, the frequency of most of the ambient vibration in practical cases is distributed over a wide frequency range from 20 to 250 Hz. Also, the broad bandgaps for in-plane and out-of-plane plate modes can hardly be overlapped with each other in same frequency range due to the weak coupling between the resonator mode and the plate mode, and thus the broad complete bandgaps are difficult to obtain too. At the same time, a common feature of phononic-crystal plates is matrix structures made of nonmetallic materials. However, most mechanical structures typically consist of metals, and a phononic-crystal plate with a nonmetallic material matrix can hardly be used for vibration and noise reduction in mechanical engineering. Recently, a metal-matrix embedded phononic crystal which consists of periodic double-sided stepped resonators deposited on a two-dimensional phononic plate within a steel matrix was proposed. It is found that the bandwidth is increased, but the opening location of the bandgap is higher (1000 Hz) [19]. In other words, the extension of the bandwidth into the lower frequency range remains a challenging task for a metal-matrix embedded phononic crystal.

Thus, how to adjust the in-plane and out-of-plane gaps overlapping with each other in the lower frequencies (below 100 Hz) or increasing the in-plane and out-of-plane gaps in the same range is an important issue for the control of the vibration in a practical case.

In this chapter, two kinds of metal-matrix embedded phononic crystals are introduced. The one is called the lower frequency complete bandgap metal-matrix embedded phononic crystals [14] because it can produce lower frequency complete bandgap, and the other is called the broad complete bandgap metal-matrix embedded phononic crystals [20] because it can produce broad complete bandgap.

2. The method of calculation

In order to investigate the band properties of the metal-matrix embedded phononic crystals, a series of calculations on dispersion relations and transmission spectra are conducted with FEM based on the Bloch theorem. For the calculation of the dispersion relations, the governing field equations for elastic wave propagation in solids are given by

$$\sum_{j=1}^{3}\frac{\partial}{\partial x_j}\left(\sum_{l=1}^{3}\sum_{k=1}^{3}c_{ijkl}\frac{\partial u_k}{\partial x_l}\right) = \rho\frac{\partial^2 u_i}{\partial t^2}\ (i = 1, 2, 3) \tag{1}$$

where ρ is the mass density, t is the time, u_i is the displacement, c_{ijkl} are the elastic constants, and x_j (j = 1, 2, 3) represents the coordinate variables x, y, and z, respectively. According to the Bloch theorem, in the FEM formulation, the displacement field can be expressed as

$$u(r) = e^{i(k\cdot r)}u_k(r) \tag{2}$$

where u is the displacement at the nodes and r is the position vector located at the boundary nodes. The Bloch wave vector k = (k_x, k_y) is the wave vector limited to the first Brillouin zone of the reciprocal lattice. Since the metal-matrix embedded phononic crystal plate is periodic in the xy-direction and finite in the z-direction, only the unit cell (shown as in **Figure 1(b)** or **Figure 3(b)**) needs to be considered in the FEM calculation when the Bloch theorem is adopted on the boundaries between the unit cell and its two adjacent cells, given by

$$u_i(x + a, y + a) = e^{i(k_x\cdot a + k_y\cdot a)}u_i(x, y)\,(i = x, y, z) \tag{3}$$

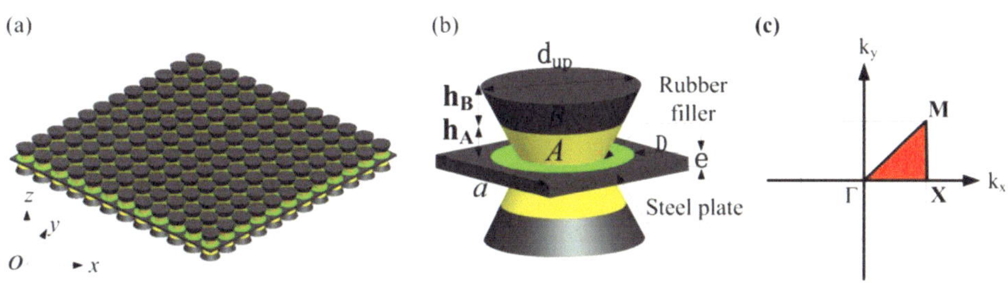

Figure 1.
(a and b) Schematic of the part and the unit cell of the lower frequency complete bandgap metal-matrix embedded phononic crystals, respectively. (c) The corresponding first irreducible Brillouin zone (red region) and the high-symmetry M, Γ, and X.

where the elastic displacement vector is denoted by u; the position vectors are denoted by x, y, and z; and k_x and k_y are the Bloch wave vectors limited in the irreducible first Brillouin zone (shown as in **Figure 1(c)** or **Figure 3(c)**). The Bloch calculation gives the eigenfrequencies and the corresponding eigenvectors, and then the dispersion relationships can be obtained by changing the wave vector in the first irreducible Brillouin zone.

To further demonstrate the existence of the bandgaps of the metal-matrix embedded phononic crystals, the transmission spectra for a structure with finite units along the x- or y-direction is calculated by using FEM. The acceleration excitation source is incident from one side (left side) of the finite structure and propagates along the x- or y-direction. The corresponding transmitted acceleration is recorded on another side (right side) of the structure. The transmission spectrum is defined as

$$TL = 10\log\left(\frac{\alpha_o}{\alpha_i}\right) \quad (4)$$

where α_o and α_i are the output and input accelerations of the metal-matrix embedded phononic crystals, respectively. Finally, the transmission spectra can be obtained by changing the excitation frequency of the incident acceleration.

3. The model and the results

3.1 The lower frequency complete BG phononic crystals

3.1.1 The model of the phononic crystals

The lower frequency bandgap metal-matrix embedded phononic crystal [14] was composed of a square array of composite taper stubs on both sides of a two-dimensional binary locally resonant PC plate which composes an array of rubber fillers embedded in the steel plate. **Figure 1(a)** and **(b)** shows part of the proposed structure and its unit cell, respectively.

In the lower frequency bandgap metal-matrix embedded phononic crystals, the taper stub is composed of the A taper cap and the B which is located on the taper A. The geometrical parameters of the structure are defined as follows: the diameter of the rubber filler, the steel plate thickness, and the lattice constant are denoted by D, e, and a, respectively; the height and the diameter of the taper stub are denoted by h (h_A for taper A and h_B for taper B) and d (the upper diameter of taper stub is denoted by d_{up}, and the lower diameter is denoted by d_{low}), respectively. The material parameters used in the calculations are listed in **Table 1**. The taper A and the taper B are rubber and steel, respectively.

3.1.2 The results of the phononic crystals

Through the use of the finite element method, the systems described in **Figure 1** were studied numerically. The band structures and displacement vector

Material	Mass density (kg/m^3)	Young's modulus (10^6N/m^2)	Poisson's ratio
Steel	7800	210,000	0.29
Rubber	1300	0.1175	0.47

Table 1.
Material parameters in calculations.

fields were computed according to the Bloch theorem. The single-unit cell (as shown in **Figure 3(b)**) is determined by the periodicity of the structure. The following structure parameters are used: D = 8 mm, e = 1 mm, a = 10 mm, h = 5 mm ($h_A = h_B = 2.5$ mm), $d_{up} = 9$ mm, and $d_{low} = 5$ mm, respectively.

It can be observed that there are 13 bands within 0–200 Hz in **Figure 2(a)**. Besides the traditional plate modes, which are the in-plane modes (mainly the symmetric Lamb modes, such as modes S_2) and the out-of-plane modes (mainly the antisymmetric Lamb modes, such as mode A_2), lots of flat modes (such as modes S_1, A_1, F_2), which are the resonant modes of the composite taper stubs, can be found. The bandgaps (one in-plane bandgap, one out-of-plane bandgap, and one complete bandgap), as a result of the coupling of the two kinds of modes mentioned above, appear. The in-plane bandgap (blue-dashed area: the frequency bands in which no in-plane modes) is due to the coupling between the in-plane modes (modes S_2) and the corresponding flat modes (modes S_1). It ranges from 53 to 93 Hz (between the fifth and eighth bands). The absolute bandwidth of it is 40 Hz. The out-of-plane bandgap (green-dashed area: the frequency bands in which no out-of-plane modes) is due to the coupling between the out-of-plane modes (mode A_2) and the corresponding flat modes (mode A_1). It ranges from 59 to 154 Hz (between the sixth and ninth bands), and the absolute bandwidth is 95 Hz; the complete bandgap (red-dashed area: the frequency bands in which neither in-plane modes nor out-of-plane modes) is due to the overlap between the in-plane bandgap and the out-of-plane bandgap. It ranges from 59 to 93 Hz (between the sixth and eighth bands). The absolute bandwidth of it is 34 Hz. It can be observed that the location of the bandgap shifts into lower frequency (below 100 Hz), but the bandwidth is very narrow.

As a comparison, we also calculated the band structures of the transition PC plate composed of double-sided composite taper stubs deposited on a homogeneous steel plate and the classical PC plate which was proposed by Assouar [9]. They are shown in **Figure 2(b)** and **(c)**, respectively. Their complete bandgaps (red-dashed areas) are both due to the overlap between the second in-plane bandgap and the first out-of-plane bandgap. It can be found clearly that the introduction of the

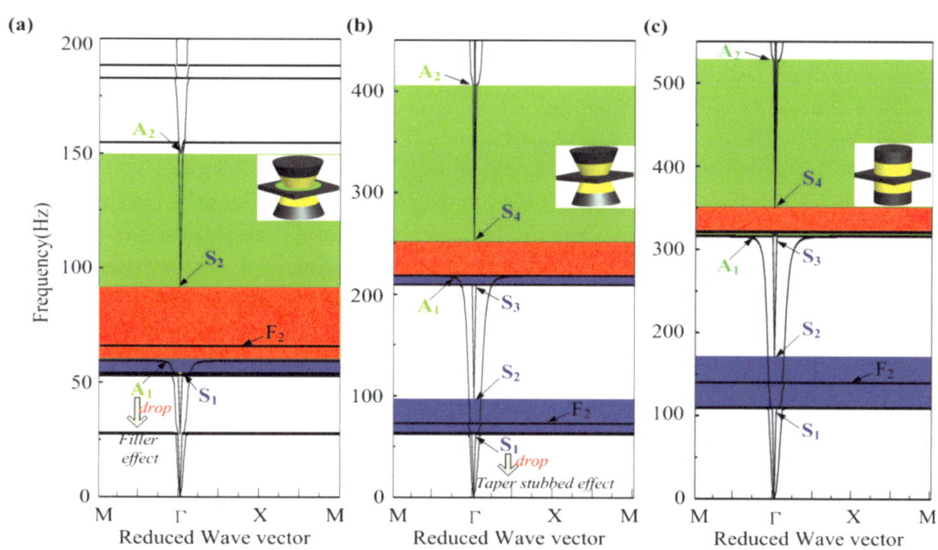

Figure 2.
Band structures of (a) the lower frequency complete bandgap metal-matrix embedded PC plate, (b) the transition PC plate, and (c) the classical PC plate. The insets are the schematic view of the unit cell of the corresponding structure. The red, blue, and yellow shadow regions denote the complete, in-plane, and out-of-plane bandgaps, respectively.

proposed structure gives rise to a significant lowering of the opening location of the first complete bandgap by a factor of 5.5 compared with the classical PC plate. Compared with the classical PC plate, introducing the double-sided composite taper stubs, the locations of both the in-plane and out-of-plane bandgaps are lowered, but the out-of-plane bandgap is always overlapped with the second in-plane bandgap; when introducing the rubber filler, the location of the in-plane bandgap is kept stationary, and the out-of-plane bandgap is shifted to lower frequency (59 Hz) overlapped with the first in-plane bandgap. Finally, a complete bandgap is generated in lower frequency (below 100 Hz). Therefore, the double-sided taper stub has a direct effect on the lowering of the location of the in-plane bandgaps (53 Hz), and the rubber filler has a direct effect on the lowering of the location of the out-of-plane bandgaps (59 Hz). It makes the out-of-plane bandgaps overlap with the first in-plane bandgap and leads to a complete bandgap in lower frequencies [14].

3.2 The broad complete BG phononic crystals

3.2.1 The model of the phononic crystals

The broad complete bandgap metal-matrix embedded phononic crystal [18] was composed of a square array of single "hard" cylinder stubs on both sides of a two-dimensional binary locally resonant PC plate which composes of an array of rubber fillers embedded in the steel plate. **Figure 3(a)** and **(b)** shows part of the structure and its unit cell, respectively.

The metal-matrix embedded phononic crystals with single "hard" cylinder stubs which consist of "hard" stub such as steel stubs contact with rubber filler. The geometrical parameters of the structure are defined as follows: the diameter of the rubber filler, the steel plate thickness, and the lattice constant are denoted by D, e, and a, respectively; the height and the diameter of the stub are denoted by h and d, respectively. The material parameters used in the calculations are listed in **Table 1**.

3.2.2 The results of the phononic crystals

Through the use of the finite element method, the systems described in **Figure 3** were studied numerically. The band structures and displacement vector fields were computed according to the Bloch theorem. The single-unit cell (as shown in **Figure 3(b)**) is determined by the periodicity of the structure. The following structure parameters are used: D = 8 mm, e = 1 mm, a = 10 mm, h = 2.5 mm, and d = 7.5 mm.

There are 12 bands within 0–600 Hz (see **Figure 4(a)**). Besides the traditional plate modes, which are the in-plane modes (such as mode S_2) and the out-of-plane modes (such as mode A_2), many flat modes (such as modes S_1, A_1, F_1), which are

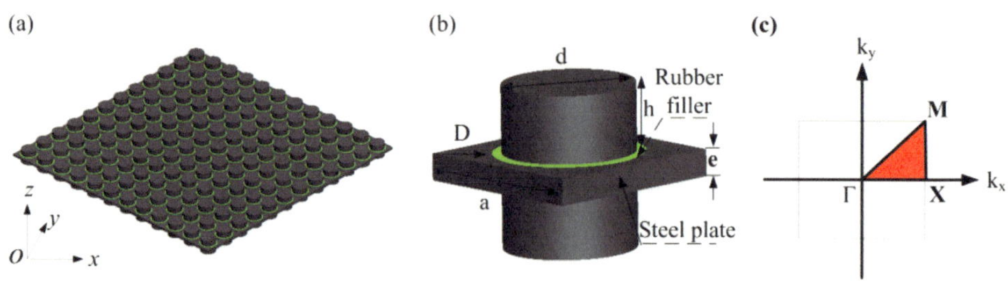

Figure 3.
(a and b) Schematic of the part and the unit cell of the broad complete bandgap metal-matrix embedded phononic crystals, respectively. (c) The corresponding first irreducible Brillouin zone (red region) and the high-symmetry M, Γ, and X.

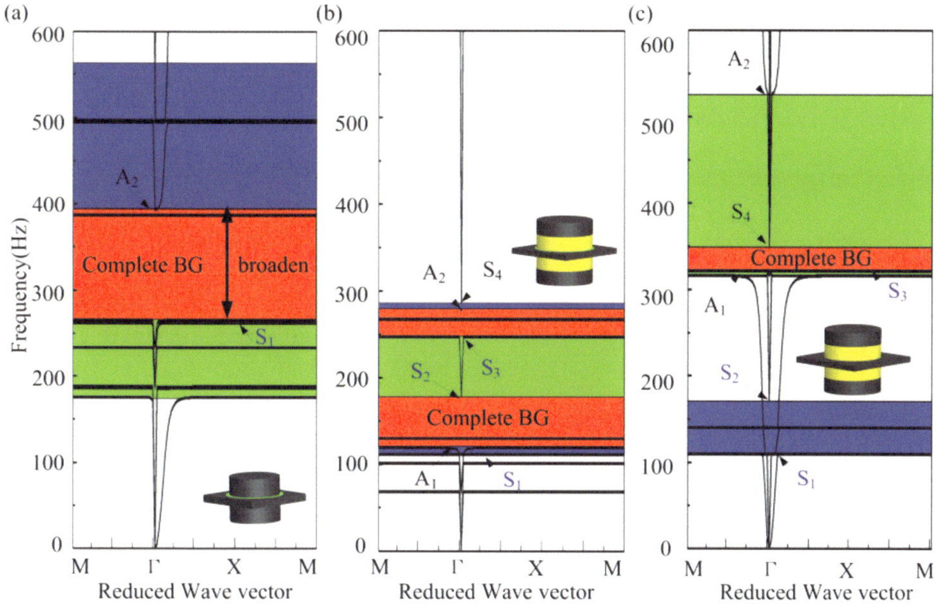

Figure 4.
Band structures of (a) the broad complete bandgap metal-matrix embedded phononic crystals, (b) the transition PC plate, and (c) the classical PC plate. The insets are the schematic view of the unit cell of the corresponding structure. The red, blue, and yellow shadow regions denote the complete, in-plane, and out-of-plane bandgaps, respectively.

the resonant modes of the single "hard" cylinder stubs, can be found. The bandgaps (one in-plane bandgap, one out-of-plane bandgap, and one complete bandgap) appear due to coupling between the two modes. The in-plane bandgap (**Figure 4(a)**: blue-dashed area) is due to coupling between the in-plane modes S_2 and the corresponding flat mode S_1. It ranges from 260 to 573 Hz (between the sixth and twelfth bands). The absolute bandwidth is 313 Hz. The out-of-plane bandgap (**Figure 4(a)**: green-dashed area) is due to coupling between the out-of-plane mode A_2 and the corresponding flat modes A_1. It ranges from 187 to 396 Hz (between the third and eighth bands), and the absolute bandwidth is 209 Hz; the complete bandgap (**Figure 4(a)**: red-dashed area) is due to the overlap between the in-plane bandgap and the out-of-plane bandgap. It ranges from 260 to 396 Hz (between the sixth and eighth bands). The absolute bandwidth is 136 Hz.

As a comparison, we also calculated the band structures of the transition PC plate composed of double-sided composite cylinder stubs deposited on a two-dimensional locally resonant phononic crystal plate that consists of an array of rubber fillers embedded in a steel plate and the classical PC plate which was proposed by Assouar [9]. They are shown in **Figure 4(b)** and **(c)**, respectively. For the classical phononic-crystal plate, its first complete bandgap (red-dashed area) is caused by the overlap between the second in-plane bandgap and the first out-of-plane bandgap. The associated absolute bandwidth is 29 Hz. For the transition phononic-crystal plate, the bandgaps are lowered after introducing the rubber filler, such that both the out-of-plane bandgap and in-plane bandgap were lowered. However, the out-of-plane bandgap is lowered more and overlaps with two in-plane bandgaps (the first and the second in-plane bandgaps). This causes the complete bandgaps to be increased, but the absolute bandwidth is also narrow (78 Hz). These phenomena confirm that the single "hard" cylinder stub has a special effect on the bandwidth and the bandwidth can be increased by introducing it. This occurs mainly because the in-plane bandgap is increased by introducing the single "hard" cylinder stub in the broad complete bandgap metal-matrix embedded phononic crystals. The absolute bandwidth of the in-plane bandgap is increased by a factor of 3.37 compared with a classical phononic-crystal plate [18].

4. Forming mechanisms of the BGs of the phononic crystals

4.1 Forming mechanisms of the lower frequency complete BGs

In order to study the physical mechanism for the occurrence of the lower frequency complete bandgap in the metal-matrix embedded phononic crystals [14, 18], several specific resonance modes (mode A_1, mode S_1), which correspond to the lower edge of the first bandgap (out-of-plane and in-plane bandgap), and several specific traditional plate modes (mode A_2, mode S_2), which correspond to the upper edge of the first bandgap (in-plane and out-of-plane bandgaps), are extracted.

4.1.1 Forming mechanisms of the lower out-of-plane BGs

Figure 5 displays the magnitude of the total displacement vector of a unit cell of the lower frequency complete bandgap metal-matrix embedded phononic crystals (**Figure 2(a)**), the transition structure (**Figure 2(b)**), and the classical structure (**Figure 2(c)**), respectively. They correspond to the upper and lower edge of the out-of-plane bandgap of each structure. The mode A_2 is an antisymmetric Lamb mode of the plate. The steel plate vibrates along the z-axis, while the stub remains stationary. In the frequencies, the antisymmetric Lamb mode will be activated, and the out-of-plane waves propagate through the PC plate in the antisymmetric Lamb mode. When the frequency of the out-of-plane waves is near the first nature frequency of the stub resonator, the resonant mode A_1 will be activated. The stub vibrates along the z-direction, and it gives a reacting force to the plate against the plate that vibrates along the z-direction. In that case, the out-of-plane waves are not capable of propagating through the PC plate. As a result, an out-of-plane bandgap is opened. Within the out-of-plane bandgap, the reacting force is still applied on the

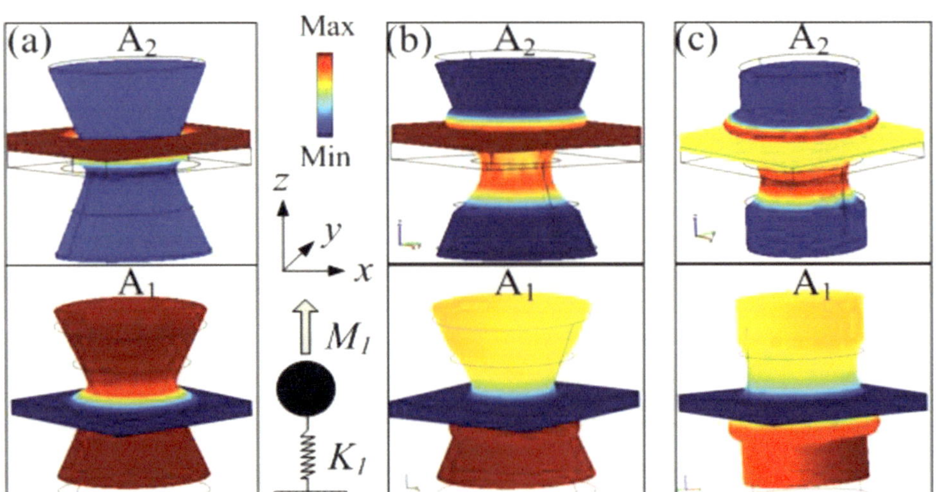

Figure 5.
The total displacement vector fields of the modes (resonant mode A_1 and antisymmetric Lamb mode A_2) (a) correspond to **Figure 2(a)**, *(b) correspond to* **Figure 2(b)**, *and (c) correspond to* **Figure 2(c)**.

plate and prevents the propagation of the out-of-plane waves. The frequency of the out-of-plane waves deviates from the nature frequency of the mode A_1, according to the modal superposition principle which can be described as

$$z = \eta_1 A_1 + \eta_2 A_2 + \cdots + \eta_n A_n \tag{5}$$

where z denotes the response of the plate and η_n denotes the modal participation factor of the mode A_n. The modal participation factor η_1 of the mode A_1 becomes small, leads the reacting force to becoming weak, and then disappears, and the antisymmetric Lamb mode A_2 is released again. As a result, the out-of-plane bandgap is closed. The formation mechanism of the bandgaps is shown in **Figure 6**. As the frequency of the out-of-plane waves deviates from the nature frequency of the resonator mode, the modal participation factors of it become small, lead the reacting force to becoming weak, and then disappear; the out-of-plane bandgap is closed. It can be concluded that the out-of-plane bandgap of the system is formed due to the coupling between the flat mode A_1 and the antisymmetric Lamb mode A_2 which is established based on the modal superposition principle.

The opening location of the out-of-plane bandgap is determined by the nature frequency of the resonant mode A_1. The vibration process of the resonant mode A_1 can be understood as a mass-spring system whose frequency is determined by the formula

$$f_1 = \frac{1}{2\pi}\sqrt{\frac{K_1}{M_1}} \tag{6}$$

where K_1 is the spring stiffness and M_1 is the lump mass. For the classical structure, the rubber stub (denoted by A) acts as a spring, and the cap steel stub (denoted by B) acts as a mass ($M_1 = M_B$, where M_B denotes the mass of stub B). For the lower frequency complete bandgap metal-matrix embedded phononic crystals, it can be found that the displacement fields are distributed in the whole stub and manifest an "analogous-rigid mode" of the whole stub, since the whole stub bodily moves along the z-axis with weak constrain, and the natural frequency is not zero. In this case, the rubber filler acts as a spring, and the whole stub acts as a mass; thus, the frequency is shifted to a lowest frequency range. It can be concluded that the out-of-plane bandgap is adjusted into lower frequency range by the rubber filler.

Figure 6.
The formation mechanism of the out-of-plane bandgaps of the lower frequency complete bandgap metal-matrix embedded phononic crystals.

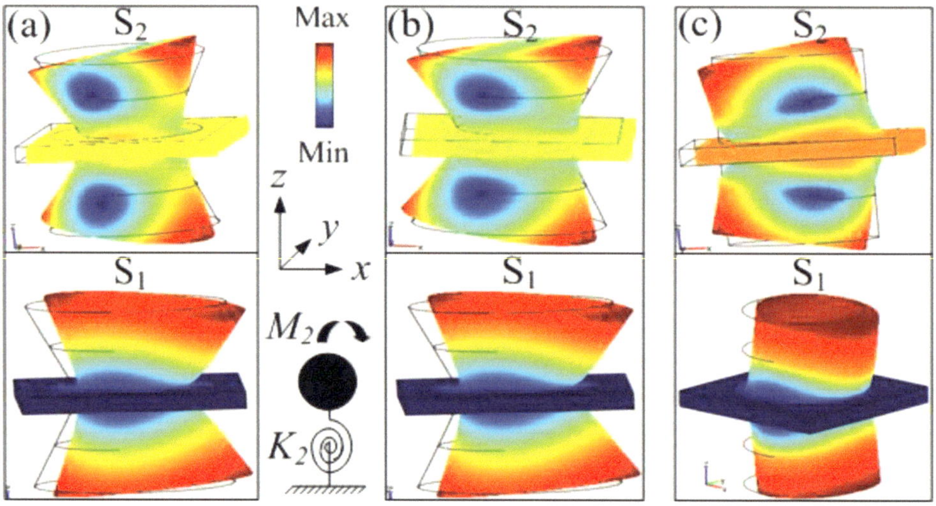

Figure 7.
The total displacement vector fields of the modes (resonant mode S_1, symmetric Lamb mode S_2) (a) correspond to ***Figure 2(a)****, (b) correspond to* ***Figure 2(b)****, and (c) correspond to* ***Figure 2(c)****.*

4.1.2 Forming mechanisms of the lower in-plane BGs

Figure 7 displays the magnitudes of the total displacement vectors of a unit cell of the lower frequency complete bandgap metal-matrix embedded phononic crystals (**Figure** 7(**a**)), the transition structure (**Figure** 7(**b**)), and the classical structure (**Figure** 7(**c**)), respectively. They correspond to the upper and lower edges of the first in-plane bandgap of each structure. The mode S_2 is a symmetric Lamb mode of the plate. The steel plate vibrates along the xy-plane, while the stubs "swing" in the opposite direction. In the frequencies, the symmetric Lamb mode S_2 will be activated, and the in-plane waves propagate through the PC plate in the symmetric Lamb mode. When the frequency of the in-plane waves is near the first nature frequency of stub resonator, the resonant mode S_1 (flat mode) will be activated. The stub "swings" along a plane which is vertical to the xy-plane, and it gives a reacting force to the plate to prevent the plate to vibrate along the xy-plane. In that case, the in-plane waves are not capable of propagating through the PC plate. As a result, an in-plane bandgap is opened. Within the in-plane bandgap, the reacting force is still applied on the plate and prevents in-plane waves that propagate. The frequency of the in-plane waves deviates from the nature frequency of the resonant mode S_1, according to the modal superposition principle. Which can be described as

$$xy = \eta_1 S_1 + \eta_2 S_2 + \cdots + \eta_n S_n \tag{7}$$

where xy denotes the response of the plate and η_n denotes the modal participation factor of the mode S_n. The modal participation factor η_1 of the mode S_1 becomes small, leads the reacting force to becoming weak, and then disappears, and then the symmetric Lamb mode S_2 is released again. As a result, an in-plane bandgap is closed. It can be concluded that the first in-plane bandgap of the system is formed due to the coupling between the flat mode S_1 and the symmetric Lamb mode S_2 which is established according to the modal superposition principle. The opening location is determined by the nature frequency of the resonant mode S_1. The resonant mode S_1 can also be understood as a "mass-spring" system whose frequency is determined by the formula

$$f_2 = \frac{1}{2\pi}\sqrt{\frac{K_2}{M_2}} \tag{8}$$

where K_2 is the tensional stiffness of the spring and M_2 is the lump mass. For the three PC plates, the rubber stub A mainly acts as a spring, and the cap steel stub B acts as a mass ($M_2 = M_B$). As the stiffness of the taper rubber stub is weaker than that of the cylinder stub, the location of the in-plane bandgap is adjusted into lower frequency range.

4.1.3 Forming mechanisms of the lower complete BGs

We can conclude from the above investigations that the resonator in the metal-matrix embedded phononic crystals can be considered as a spring-mass system, as shown in **Figures 6** and 7. The spring-mass system includes two subsystems. The first one is the K_1-M_1 subsystem (also refer to **Figure 6**). Its local resonance mode is coupled with the antisymmetric Lamb mode of the plate according to the modal superposition principle, and then the out-of-plane bandgap is generated. The other one is the K_2-M_2 subsystem (also **Figure 7**). Its local resonance mode is coupled with the symmetric Lamb mode of the plate according to the modal superposition principle and is responsible for the formation of the in-plane bandgaps. The opening locations of the in-plane and out-of-plane BGs depend on the natural frequencies of the two subsystems.

As for the classical and the transition metal-matrix embedded phononic crystals, the steel stub (stub B) acts as a mass, and the rubber stub (stub A) acts as a spring, which leads to the coupling between the K_1-M_1 subsystem and K_2-M_2 subsystem. Therefore, it is difficult to adjust the in-plane and out-of-plane bandgaps separately. In the lower frequency bandgap metal-matrix embedded phononic crystals, the rubber filler acts as the stiffness K_1, the stub A acts as the stiffness K_2, the whole stub acts as the mass M_1, and the stub B acts as the mass M_2. As a result, the coupling between the K_1-M_1 subsystem and the K_2-M_2 subsystem can be decoupled. Moreover, the mass M_1 is magnified due to the "analogous-rigid mode" of the whole stub. Therefore, the out-of-plane bandgap can be adjusted into the lowest frequency range. Additionally, the stiffness K_2 can be reduced by introducing the taper stub, and thus the in-plane bandgaps can be adjusted into the lowest frequency range. As a result, the two bandgaps can be overlapped with each other in the lowest frequency, and finally a lowest complete bandgap is formed [14].

4.2 Forming mechanisms of the broad complete BGs

In order to study the physical mechanism for the occurrence of a broad complete bandgap in the metal-matrix embedded phononic crystals, several specific resonance modes (modes A_1 and S_1), which correspond to the lower edge of the first bandgap (out-of-plane and in-plane bandgap), and several specific traditional plate modes (mode A_2 and S_2), which correspond to the upper edge of the first bandgap (in-plane and out-of-plane bandgaps), are extracted.

4.2.1 Forming mechanisms of the broad out-of-plane BGs

Figure 8 displays the magnitude of the total displacement vector of a unit cell of the broad complete bandgap metal-matrix embedded phononic crystals (**Figure 8(a)**), the transition PC plate (**Figure 8(b)**), and the classical PC plate (**Figure 8(c)**), respectively. They correspond to the upper and lower edge of the out-of-plane bandgap of each structure. The mode A_2 is an antisymmetric Lamb mode of the plate. The steel plate vibrates along the z-axis, while the stub remains

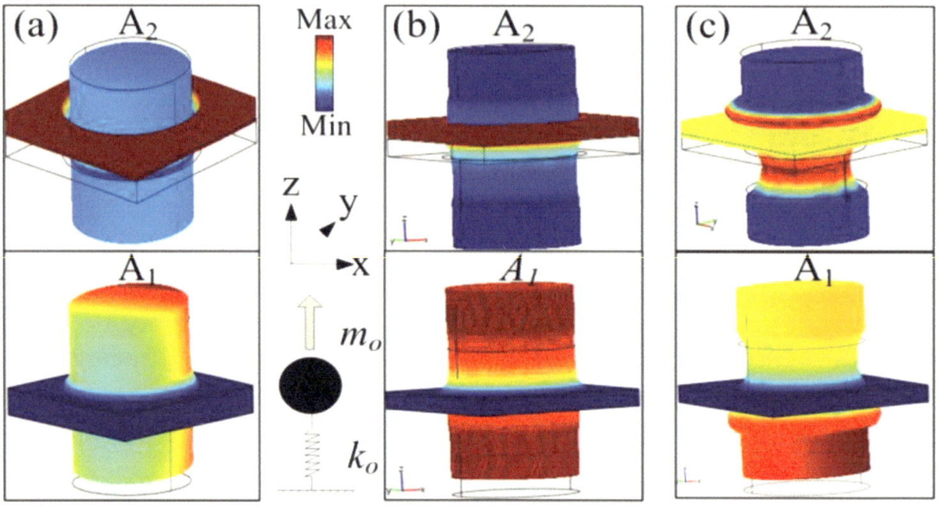

Figure 8.
*The total displacement vector fields of the modes (resonant mode A1 and antisymmetric Lamb mode A2) (a) correspond to **Figure 4(b)**, (b) correspond to **Figure 4(b)**, and (c) correspond to **Figure 4(c)**.*

stationary. With regard to the frequencies, the antisymmetric Lamb mode will be activated, and the out-of-plane waves propagate through the phononic-crystal plate in the antisymmetric Lamb mode. When the frequency of the out-of-plane waves is near the first nature frequency of the stub resonator, the resonant mode A_1 will be activated. The stub vibrates along the z-direction and produces a reacting force for the plate, and the plate vibrates along the z-direction. In this case, the out-of-plane waves are not capable of propagating through the phononic-crystal plate. As a result, an out-of-plane bandgap is created. The bandwidth is determined by the coupling between the resonance mode A_1 and the traditional plate A_2 mode.

The coupling strength between the resonance mode A_1 and the traditional plate A_2 mode is determined by the formula

$$F = k_o s_T \tag{9}$$

where s_T is the vibration amplitude of the stub and k_o is the spring stiffness.

For the classical structure, the formation mechanism of the out-of-plane bandgap is shown in **Figure 9(b)**. The hard stub (rigid body) vibrates along the z-direction and generates a reacting force through the soft stub (flexible body) to the plate against the plate vibrating along the z-direction. The soft stub acts as a spring, while the hard stub acts as a mass ($k_o = k_{sc}$, $m_o = m_h$, where m_h denotes the mass of the hard stub and k_{sc} denotes the compression stiffness of the soft stub) (see **Figure 10(b)**).

For both the broad complete bandgap metal-matrix embedded phononic crystals and the transition PC plate, it can be found that the displacement fields are distributed in the whole stub (see **Figure 8(a)** and **(b)**), respectively. This results in an "out-of-plane analogous-rigid mode" because the whole stub vibrates along the z-axis (out-of-plane) with a weak constraint, while the frequency is non-zero. The displacement fields of its eigenmodes are distributed throughout the whole stub. This means that the whole stub body moves along the z-direction like a rigid body moves in rigid mode. However, the natural frequency is not 0, and the whole stub is constrained by the rubber filler. Therefore, we refer to the concept "rigid mode" and call these types of vibration modes for the whole stub the "out-of-plane analogous-rigid mode." The formation mechanisms for the out-of-plane bandgap of the two structures are shown in **Figure 9(a)** and **(c)**, respectively.

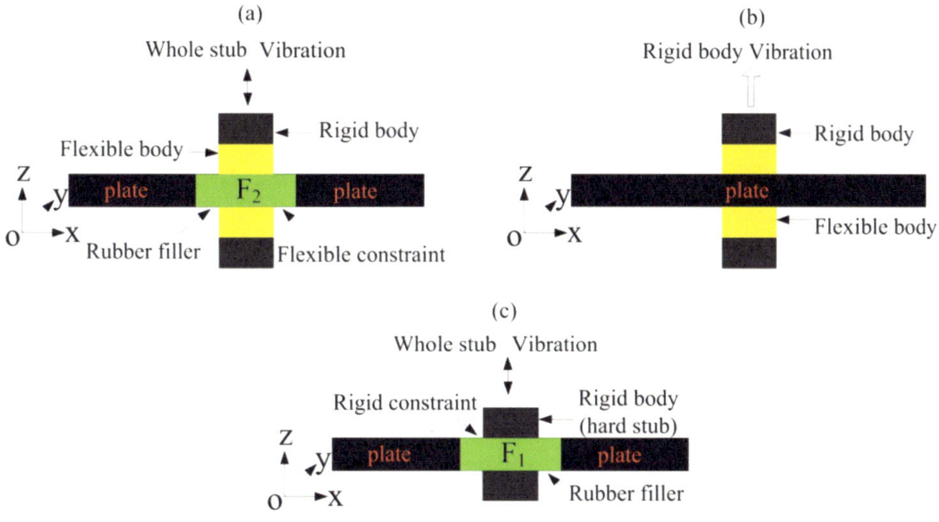

Figure 9.
Formation mechanism of the bandgap for the out-of-plane bandgap of (a) the transition PC plate, (b) the classical PC plate, and (c) the broad complete bandgap metal-matrix embedded phononic crystals.

The whole stub vibrates in the z-direction (out-of-plane) and generates a reacting force. F_1 and F_2 respond to the broad complete bandgap metal-matrix embedded phononic crystals and the transition PC plate, respectively. $F_1 = k_0.s_1 = k_{RT}.s_1$, where k_{RT} is the transverse stiffness of the rubber filler and s_1 is the vibration amplitude of the whole single "hard" stub; $F_2 = k_0.s_2 = k_{RT}.s_2$, where s_2 is the vibration amplitude of the whole composite stub through rubber filler to the plate against the plate vibrates along the z-direction (see **Figure 9(a)** and **(c)**).

In this case, the rubber filler acts as a spring, and the whole stub acts as a mass. The broad complete bandgap metal-matrix embedded phononic crystals

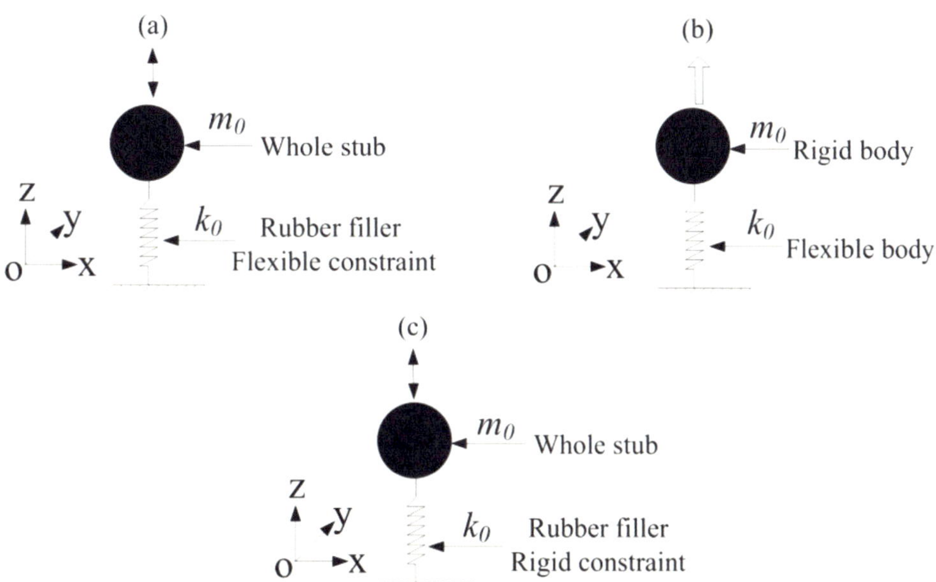

Figure 10.
The equivalent theoretical model of the resonator for formation mechanism of the out-of-plane bandgap of (a) the transition PC plate, (b) the classical PC plate, and (c) the broad complete bandgap metal-matrix embedded phononic crystals.

includes $m_o = 2m_h$ and $m_o = 2m_s + 2m_h$, where m_s denotes the mass of soft stub and m_h denotes the mass of hard stub. It is shown in **Figures 9(c)** and **10(c)** that the frequency is shifted to the lowest frequency range. Compared to the transition PC plate, where the soft stub contacts the rubber filler such that it represents a flexible constraint to the rubber filler (see **Figure 9(a)**), the hard stub contacts the rubber filler in the broad complete bandgap metal-matrix embedded phononic crystals to produce a rigid constraint for the rubber filler (see **Figure 9(c)**). This causes the spring stiffness k_0 ($k_0 = k_{RT}$—the longitudinal stiffness of the rubber filler) to increase, while the lump mass becomes smaller ($2m_s + 2m_h > 2m_h$). This, in turn, not only causes the opening location of the out-of-plane bandgap to shift to higher frequencies but also makes the force F_1 larger than the force F_2. As a result, the out-of-plane bandwidth becomes wider. We conclude that, after introducing the rubber filler, an out-of-plane analogous-rigid mode of the stub was produced, which can reduce the location of the out-of-plane bandgap. Hence, the introduction of a single "hard" stub increases the bandwidth of the out-of-plane bandgap by enhancing the stiffness of the out-of-plane analogous-rigid mode of the stub.

4.2.2 Forming mechanisms of the broad in-plane BGs

Figure 11 displays the magnitudes of the total displacement vectors of a unit cell of the broad complete bandgap metal-matrix embedded phononic crystals, the transition PC plate, and the classical PC plate. The figures correspond to the upper and lower edges of the first in-plane bandgap of each structure. The mode S_2 is a symmetric Lamb mode of the plate. The steel plate vibrates along the xy-plane (in-plane), while the stubs swing in the opposite direction. With respect to the frequencies, the symmetric Lamb mode will be activated, and the in-plane waves can propagate through the phononic-crystal plate in the symmetric Lamb mode. When the frequency of the in-plane waves approaches the first natural frequency of the stub resonator, the resonant mode S_1 will be activated. The stub vibrates in the yz-plane and produces a reacting force to the plate while vibrating in the xy-direction. In that case, the in-plane waves are not able to propagate through the PCs. As a result, an in-plane bandgap opens up. The bandwidth is determined by the coupling between the resonance mode S_1 and the traditional plate mode S_2.

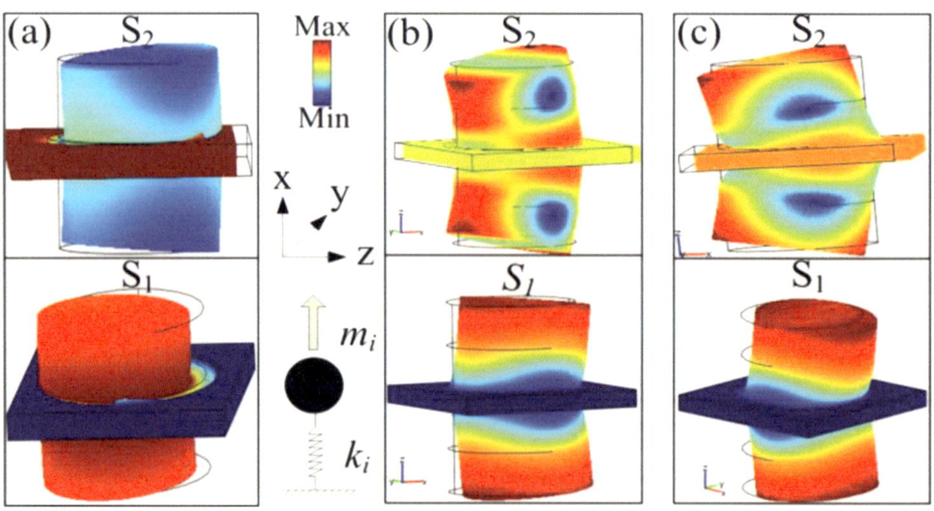

Figure 11.
*The total displacement vector fields for the modes (resonant mode S1, symmetric Lamb mode S2) (a) correspond to **Figure 4(a)**, (b) correspond to **Figure 4(b)**, and (c) correspond to **Figure 4(c)**.*

The coupling strength between the resonance mode S_1 and the traditional plate mode S_2 is determined by the formula

$$F = k_i s_L \tag{10}$$

where s_L is the vibration amplitude of the stub and k_i is the tensional stiffness of the spring.

For both the transition and the classical phononic-crystal plates, the formation mechanism of the in-plane bandgap is shown in **Figure 12(a)** and **(b)**, respectively.

It can be found that the stubs "swing" in the xy-plane and produce a reacting force ($F_2 = k_i.s_L = k_{ST}.s_L$, where k_{ST} is the transverse stiffness of the soft stub and s_L is the vibration amplitude of the whole stub) through the soft stub (flexible body) to the plate against the plate vibrating along the xy-plane. The soft stub (flexible) acts as a spring ($k_i = k_s$), while the hard stub (rigid body) acts as a mass ($m_i = m_h$), as shown in **Figure 13(a)** and **(b)**, respectively. Hence, both location and bandwidth of the first in-plane bandgap in these structures are the same.

For the broad complete bandgap metal-matrix embedded phononic crystals, it can be found that the displacement fields are distributed across the entire stub (**Figure 11(a)**) to cause an "in-plane analogous-rigid mode" because the whole stub vibrates along the xy-plane (in-plane) with a weak constraint, while the frequency is non-zero. It can be observed that the displacement fields of its eigenmodes are distributed throughout the whole stub. This means that the whole stub body moves in the xy-plane like a rigid body moves in rigid mode. However, the natural frequency is not 0, and the whole stub is constrained by the rubber filler. Therefore, we refer to the concept of rigid mode and call this type of vibration mode of a whole stub the "in-plane analogous-rigid mode." The formation mechanism of the bandgap is shown in **Figure 12(c)**. The whole stub vibrates in the xy-plane and produces a reacting force ($F_1 = k_i.s_L = k_{RL}.s_L$, where k_{RL} is the longitudinal stiffness of the rubber filler and s_L is the vibration amplitude of the whole stub) through the rubber filler to the plate against the plate vibrating in the xy-plane.

In this case, the "rubber filler" acts as a spring, where the longitudinal stiffness of the rubber filler acts as the spring, while the whole stub acts as a mass, where

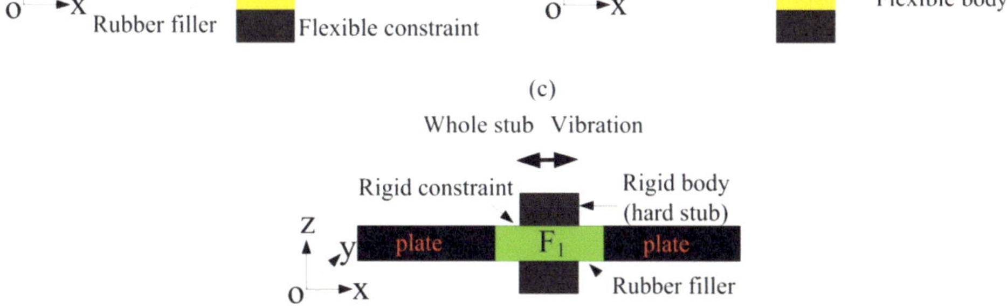

Figure 12.
Formation mechanism process of the bandgap for the in-plane bandgap of (a) the "transition" PC plate, (b) the classical PC plate, and (c) the broad complete bandgap metal-matrix embedded phononic crystals.

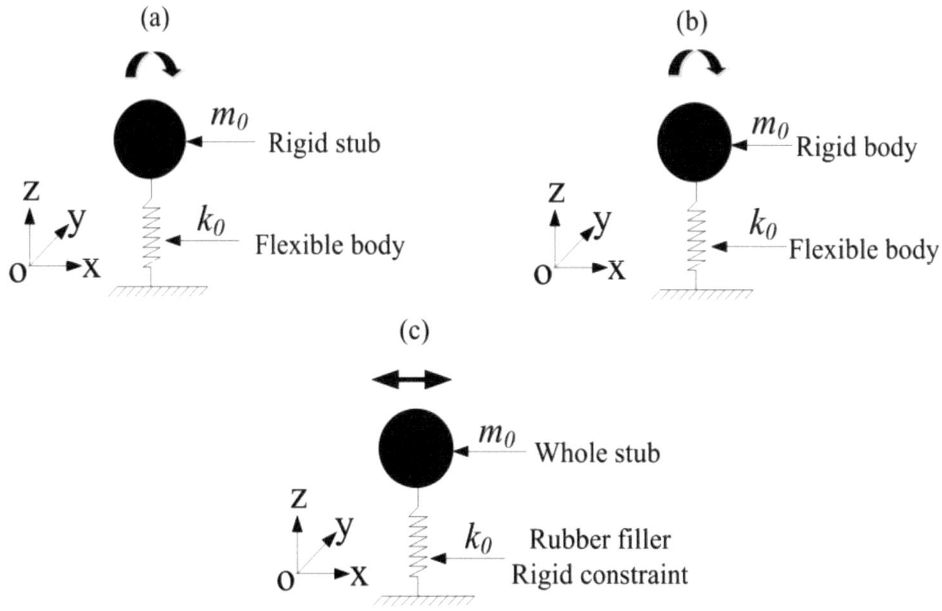

Figure 13.
The equivalent theoretical model of the resonator for formation mechanism of the in-plane bandgap of (a) the "transition" PC plate, (b) the classical PC plate, and (c) the broad complete bandgap metal-matrix embedded phononic crystals.

$m_i = m_h$, and m_h denotes the mass of hard stub (see **Figure 13(c)**). Compared to the transition PC plate, in which the soft stub contacts the rubber filler such that it gives a flexible constraint to the rubber filler, this leads to the soft stub acting as a spring (see **Figure 13(c)**). The hard stub contacts the rubber filler in the broad complete bandgap metal-matrix embedded phononic crystals, which creates a rigid constraint to the rubber filler that causes the rubber filler to act as a spring (see **Figure 13(c)**). This causes the spring stiffness k_i to increase in the broad complete bandgap metal-matrix embedded phononic crystals such that the opening location of the in-plane bandgap is shifted toward higher frequencies. However, it also makes the force F_1 larger than force F_2 and causes the in-plane bandwidth to become broader. We conclude that, after introducing the rubber filler and the single "hard" stub simultaneously, an "in-plane analogous-rigid mode" of the stub is produced, which can increase the bandwidth of the in-plane bandgap.

4.2.3 Forming mechanisms of the broad complete BGs

When the broad out-of-plane bandgaps and broad in-plane bandgaps, which were simultaneously increased by the single "hard" stub, overlap, a wider complete bandgap, in which the out-of-plane and in-plane Lamb waves are prohibited, is created.

It can be concluded that the bandwidth of the bandgap is determined by the resonator mode. After introducing the rubber filler and the single "hard" stub simultaneously, two new resonator modes are produced. One resonator mode is the in-plane analogous-rigid mode, in which the whole stub vibrates along the in-plane plate. The other resonator mode is the out-of-plane analogous-rigid mode, in which the stub vibrates along the out-of-plane plate. They both increase the in-plane and out-of-plane bandgaps, respectively. Two increased bandgaps overlap to produce a broad complete bandgap, in which both the out-of-plane and in-plane Lamb waves are prohibited. The rubber filler shifts the two kinds of bandgaps simultaneously toward low frequency such that a broad complete bandgap for low frequencies is obtained.

5. The effect of the stubs on the BGs

In order to investigate the effect of the stubs on the complete bandgaps of the metal-matrix embedded phononic crystals [14, 18], we studied the influence of the stub height on the first complete bandgap.

5.1 The effect of the stubs on the lower frequency complete BGs

Figure 14 displays the evolution of the first complete bandgap as a function of the steel-stub height, h_S. We can find that, with the increase of the steel-stub height, both the lower and upper edges of the first complete shift to a lower frequency range firstly and then move to higher frequency range. For example, when the steel-stub height is less than or equal to 3 mm, with the increase of the steel-stub height, the lower edge shifts to lower frequencies, but when the rubber stub height is larger than 3 mm, the lower edge frequency shifts to higher frequencies as the steel-stub height increases.

5.2 The effect of the stubs on the broad complete BGs

In order to investigate the effect of single "hard" stubs on the complete phononic bandgaps of the novel metal-matrix PCs, we studied the effect of the steel-stub height on the first complete bandgap. **Figure 15** displays the evolution of the first complete bandgap as a function of the steel-stub height *h*.

We find that, with the increase of the steel-stub height, the location of the first complete bandgap shifts to a lower frequency range before it shifts to a higher frequency range. Furthermore, the bandwidth of the bandgaps becomes broad. For example, when the steel-stub height is below or equal to 3.5 mm, after increasing the steel-stub height, the location of the bandgaps shifts to lower frequencies. However, when the steel-stub height exceeds 3.5 mm, the location of the bandgaps shifts to higher frequencies as the steel-stub height increases.

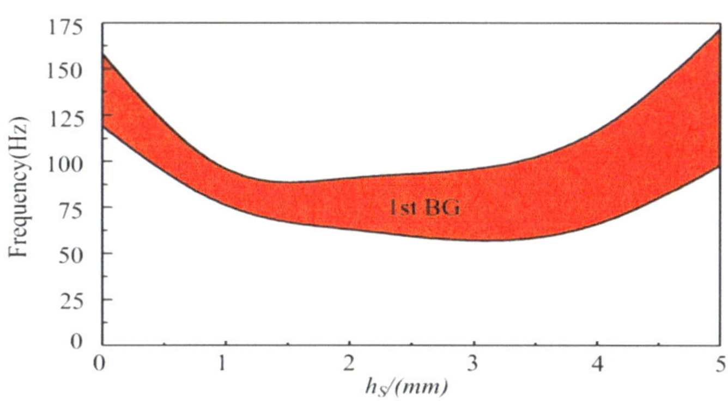

Figure 14.
The evolution of the first complete BG in the lower frequency complete bandgap metal-matrix embedded phononic crystals as a function of the steel-stub height with D = 8 mm, d_{up} = 9 mm, d_{up} = 5 mm, h = 5 mm, e = 1 mm, and a = 10 mm, respectively.

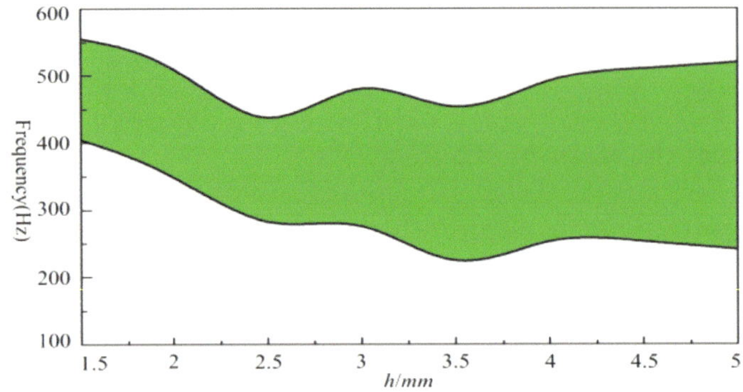

Figure 15.
The evolution of the first bandgap in the proposed phononic-crystal plate with "hard" stubs as a function of steel-stub height with D = 8 mm, d = 7.5 mm, e = 1 mm, and a = 10 mm, respectively.

6. Conclusions

In this chapter, metal-matrix embedded phononic crystals consisting of double-sided stubs (single stubs and composite stubs), which are deposited on a two-dimensional locally resonant phononic crystal plate that consists of an array of rubber fillers embedded in a steel plate, are introduced. The following summaries were drawn [14, 18]:

The spring-mass system of the resonator can be decoupled by introducing the rubber filler, and then the out-of-plane bandgap and the in-plane bandgap can be adjusted into the same lowest frequency range. The out-of-plane bandgap and the in-plane bandgap can be overlapped with each other. As a result, a lower frequency complete bandgap which ranges from 59 to 93 Hz is obtained in the metal-matrix embedded phononic crystals.

Both the out-of-plane and the in-plane phononic bandgap increase after introducing single "hard" cylinder stubs. When introducing the rubber filler and the single "hard" stub simultaneously, two new kinds of resonance modes are produced: an in-plane analogous-rigid modes, where the whole stub vibrates in-plane with the plate, and the other new resonance mode is the out-of-plane analogous-rigid mode, where the whole stub vibrates out-of-plane with respect to the plate. The out-of-plane bandgap increases for the out-of-plane analogous-rigid mode, and the in-plane bandgap increases for the in-plane analogous-rigid mode. Both the out-of-plane and the in-plane analogous-rigid modes are mainly formed due to introducing the single "hard" cylinder stub. The in-plane and out-of-plane bandgaps overlap and produce a broad complete bandgap in the metal-matrix embedded phononic crystals. Within this broad complete bandgap, both the in-plane and out-of-plane Lamb waves are prohibited. The absolute bandwidth of the bandgap for the proposed structure is five times higher than for the classic double-sided stubbed metal-matrix phononic-crystal plate.

The effect of the stub on the bandgaps is investigated. Results show that the location of the bandgaps can be modulated in a significant lower frequency range, and the bandwidth can be expanded in a considerable large frequency range by introducing different composite taper stubs, and the bandwidth can be expanded to a larger frequency range by introducing different "hard" stubs.

The proposed structure provides an effective way for phononic crystals to obtain wide complete bandgaps and lower frequency complete bandgaps (below 100 Hz), which have a potential application in the low-frequency vibration reduction in a practical case.

Acknowledgements

This chapter was supported by the Project of the National Science Foundation of China (No. 51674199).

Author details

Suobin Li[1*], Yihua Dou[1] and Linkai Niu[2]

1 School of Mechanical Engineering, Xi'an Shiyou University, Xi'an, China

2 College of Mechanical Engineering, Taiyuan University of Technology, Taiyuan, China

*Address all correspondence to: ziyedeyan@stu.xjtu.edu.cn

References

[1] Sigalas M, Economou EN. Band structure of elastic waves in two dimensional systems. Solid State Communications. 1993;**86**(3):141-143

[2] Liu Z, Mao X, et al. Locally resonant sonic materials. Science. 2000;**289**:1734

[3] Meyer R. Vibrational band structure of nanoscale phononic crystals. Physica Status Solidi. 2016;**213**(11):2927-2935

[4] Ho KM, Cheng CK, Yang Z, et al. Broadband locally resonant sonic shields. Applied Physics Letters. 2003;**83**(6):5566-5568

[5] Soliman YM, Su MF, Leseman ZC, et al. Effects of release holes on microscale solid–solid phononic crystals. Applied Physics Letters. 2010;**97**(8):081907

[6] Pennec Y, Djafari-Rouhani B, Larabi H, et al. Low-frequency gaps in a phononic crystal constituted of cylindrical dots deposited on a thin homogeneous plate. Physical Review B. 2008;**78**(10):104105

[7] Wu TT, Huang ZG, Tsai TC, Wu TC. Evidence of complete band gap and resonances in a plate with periodic stubbed surface. Applied Physics Letters. 2008;**93**(11):2022

[8] HSU J. Local resonances-induced low-frequency band gaps in two-dimensional phononic crystals slabs with periodic stepped resonators. Journal of Physics D: Applied Physics. 2011;**44**:055401

[9] Assouar MB, Oudich M. Enlargement of a locally resonant sonic band gap by using double-sided stubbed plate. Applied Physics Letters. 2012;**100**:123506

[10] Zhang SW, Wu JH, Hu ZP. Low-frequency locally resonant band-gaps in phononic crystal plates with periodic spiral resonators. Journal of Applied Physics. 2013;**113**:163511

[11] Song G, Chen J, Han X. Shear horizontal guided wave band gaps in a homogenous plate with periodic tapered surface. Japanese Journal of Applied Physics. 2014;**53**(9):94301

[12] Li Y, Chen T, Wang X, et al. Enlargement of locally resonant sonic band gap by using composite plate-type acousticmetamaterial. Physicals Letters. A. 2015;**379**:412-416

[13] Li S, Chen T, Wang X, et al. Expansion of lower-frequency locally resonant band gaps using a double-sided stubbed composite phononic crystals plate with composite stubs. Physics Letters A. 2016;**380**:2167-2172

[14] Li S, Chen T, Wang X, et al. Lamb waves propagation in a novel metal-matrix phononic crystals plate. Modern Physics Letters B. 2016;**30**:1650338

[15] Li S, Chen T, Xi Y, et al. Forming mechanisms of low-frequency complete band gaps in phononic crystal plate. Journal of Xi'an Jiaotong University. 2016;**50**(12):51-57

[16] Zhoua X, Xub Y, Liua Y, et al. Extending and lowering band gaps by multilayered locally resonant phononic crystals. Applied Acoustics. 2018;**133**:97-106

[17] Zhao H-J, Guo H-W, Gao M-X, et al. Vibration band gaps in double-vibrator pillared phononic crystal plate. Journal of Applied Physics. 2016;**119**:014903

[18] Jiang P. Low-frequency band gap and defect state characteristics in a multi-stub phononic crystal plate with slit structure. Journal of Applied Physics. 2017;**121**:015106

[19] Li Y, Zhu L, Chen T. Plate-type elastic metamaterials for low-frequency broadband elastic wave attenuation. Ultrasonics. 2017;**73**:34-42

[20] Li S, Dou Y, Chen T, et al. A novel metal-matrix phononic-crystal with a low frequency, broad and complete, locally-resonant bandgap. Modern Physics Letters B. 2018;**32**:1850221

Chapter 4

Hybrid Liquid-Crystal/Photonic-Crystal Devices: Current Research and Applications

Yu-Cheng Hsiao

Abstract

In this chapter, the current research and development of the liquid crystal-based photonic crystals is introduced. This chapter will present the essential knowledge of the new photonic crystal technology and applications in simple language. In the recent year, liquid crystal-enabled photonic crystal technologies have attracted broad attentions from scientists. Based on special optical properties of liquid crystal-enabled photonic crystal device, many applications, such as tunable optical filters, tunable optical modulators, optical pulse compressors, laser device, and applications in multiphoton microscopy, have been developed in recent years. In addition, the detailed optical properties, operation principles, and prospects are discussed in this chapter.

Keywords: photonic crystals, liquid crystals, photonic band gap, electro-optic device, optical properties

1. Introduction

1.1 Liquid crystal-based photonic crystal devices

Photonic crystals (PCs) are structural materials with periodically varying dielectric permittivity. The term PC was coined three decades ago, and fascinating properties of PCs have attracted numerous scientists to put in a great deal of effort. PCs consist of dielectric materials. In PCs, the index of refraction varies periodically in space. PCs were invented since 1987, when both Yablonovitch and John published their results independently [1, 2]. The most important characteristic of PCs is the photonic band gap (PBG). Furthermore, the PBG of PCs is an optical analog to the electronic bandgap in semiconductor materials, which means photons will be localized or forbidden in the PCs. Based on this characteristic, PCs can be used as various photonic applications [3–7]. However, if a defect layer is introduced to a PC structure, disrupting its periodicity, the transmission of photons at specific wavelengths will be induced within the PBG; these narrow transmission bands within PBG are called defect modes. Based on this special design of PCs with defect layers, many photonic device applications were proposed: PC lasers [8], PC optical fibers [9], and other optical devices [10, 11]. In addition, the spectral properties of the PCs can be controlled if the defect layer is tunable, for instance, liquid crystal (LC).

LCs are anisotropic materials. The optic axis (or molecular orientation) can be controlled by applying electric field, magnetic field, or temperature. Based on optical anisotropy, LC can be used as a phase retarder or an optical polarization rotator, of which the refractive indices can be tuned with external fields. LCs are also used in many other applications such as displays, smart windows, and optical fibers. Furthermore, by inserting a LC layer as a defect layer in PCs, the tunable defect modes can be achieved. The first tunable LC-based PC hybrid structure was developed by Ozaki et al. They employed a planar-aligned nematic LC as a central defect layer sandwiched between two one-dimensional (1D) PCs [12]. **Figure 1** shows the setup of the designed hybrid PC/LC cell. This idea was then extended to PC/cholesteric LC (CLC) structures for tunable laser applications [13–17]. CLCs are self-organized PCs, in which the molecular chirality forms helical structures with the optic axis continuously twisted along the helical axis. The effects of angle of incidence [18], temperature [19], and magnetic field [20] on the optical properties of PC/LC devices are investigated.

Zyryanov et al. investigated the hybrid PC/LC device between the crossed polarizers. Their experimental results show attractive features of the tunable defect modes within the PBG. The wavelengths of defect modes are shifted by the change in effective refractive index (n_{eff}) [21]. Larger number of defect modes can be induced by using high refractive index LC or by increasing the thickness of the LC defect layer. Moreover, they experimentally and theoretically demonstrated that the interference of defect modes can be achieved by placing the cell between crossed polarizers. The orthogonal polarization components through vector sum in the projection direction along the axis of the analyzer lead to shift of defect modes. As a result, the transmittance is increased when the defect mode wavelength of an extraordinary component overlaps that of an ordinary one in PCs [20].

In addition, the next milestone is electrically tunable photonic device based on PC/CLC and PC/polymer-stabilized CLC (PSCLC) hybrid structures [22–24]. With the PC/PSCLC structure, not only the wavelength of defect mode is switchable among multistable states by voltage pulses, but also the optical intensity of defect modes can be electrically tuned through switching among different metastable states. **Figure 2** shows the sandwiched structure of the PC/PSCLCs in the three stable states. It is wavelength switchable and intensity tunable of defect modes among three stable states. The unique optical tristability in the defect modes reduces power consumption and enhances flexibility. From then on, the PC and CLC combined

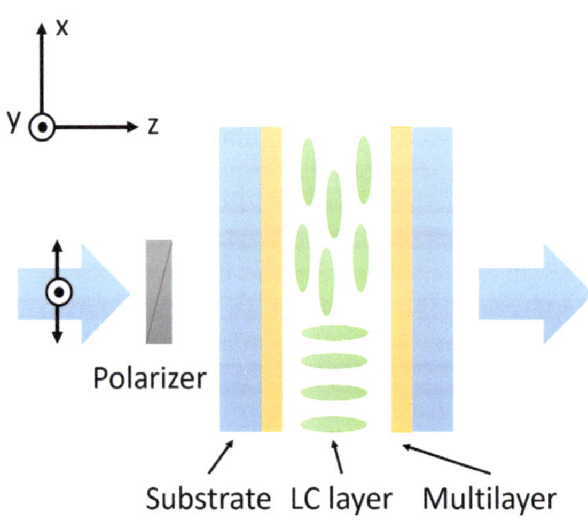

Figure 1.
Schematic of the PC/LC hybrid device in an electro-optical setup.

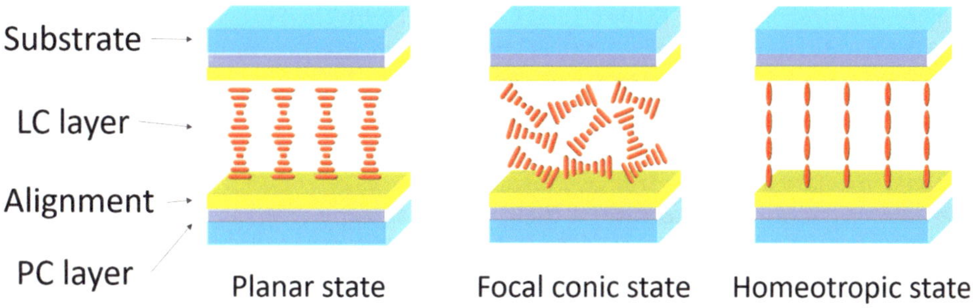

Figure 2.
Schematic of the PC/CLC hybrid device in three stable states.

devices became a hot topic. Many types of PC/chiral LC hybrid devices are invented such as PC/BHN, PC/THN, and PC/PSCT. Recently, many scientists employ hybrid PC/CLC to achieve the lasing applications [25, 26].

1.2 Aim of this chapter

An overview of the development of PC/LC is presented. The optical properties, operation principles, and applications of liquid crystal-based photonic crystal devices are discussed. In Section 1, I introduce the basic knowledge of LC and PC physics to make the reader understand the next section. Section 2 details the operation principles of nonchiral and chiral LC modes to help the understanding of switching mechanisms described in two following sections. In addition, the optical properties of defect modes in different LC states are reported in Section 2. Section 2.1 is "Photonic crystals with a nonchiral nematic liquid crystal," Section 2.2 is "Chiral-tilted homeotropic nematic liquid crystal-based photonic crystal devices," Section 2.3 is "Chiral nematic and cholesteric liquid crystal with photonic crystal devices," and Section 2.4 is "Tristable photonic crystal devices with polymer-stabilized cholesteric textures." The configuration of many types of PC/LC cell applications, including the design of PC multilayers for applications is schematically depicted in Session 3. To realize the PC applications; Session 3.1 introduces "Electrically switchable liquid crystal-based photonic crystals for a white light laser," Session 3.2 shows "Liquid crystal-based photonic crystals for pulse compression and signal enhancement in fluorescence applications," and Session 3.3 demonstrates "Photo-manipulated photonic devices based on tristable chiral-tilted homeotropic nematic liquid crystal." Finally, I will summarize the results and conclude about this chapter.

2. Operation method of liquid crystal-based photonic crystal devices

2.1 Photonic crystals with a nonchiral nematic liquid crystal

Nowadays, available LC materials can be classified into two categories: nonchiral and chiral system, according to their different operation functions on their molecule arrangement. The nonchiral LC owns only one stable arrangement state, which is determined by the use of the alignment layer. In addition, nonchiral LC molecule is continuously oriented, when the voltage is applied. On the contrary, two or multistable states are exhibited in the chiral LC system. Moreover, the multistable states can be switched and controlled from one to another based on the condition of applied voltage pulse. Based on the properties of the nonchiral LC system, it can serve as a phase retarder or optical rotator tuned by electrically controlled birefringence (ECB) effect. Combined with the PC cells from Prof. Ozaki introduced in Section 1.1, the LC defect layers used are nonchiral LCs, and it can be operated like phase retarders. In order to understand the PC/nonchiral LC, operation method of PC/LC devices is most important. The operation method of the PC/LC in nonchiral type as tunable defect can be described in terms of two effects: ordinary and extraordinary refractive indices effects [12, 20, 27]. From **Figure 1** in Section 1.1, the positive dielectric anisotropy of the nonchiral LC is aligned along the x-axis as well as the light propagates along the z-axis. In addition, the PC structure consists of two dielectric materials: high and low refractive index materials, which are stacked alternatively. When the light is incident to the hybrid PC/LC device normally as well as the light polarization direction parallel to the LC molecule along the x-axis, the extraordinary refractive index (n_e) contributed the optical path length (OPL) is exhibited. Thus, the appearance of defect modes in the PBG represents extraordinary defect modes. When the electric field is applied across the PC/LC device to make LC molecule along the z-axis, the OPL is only contributed by the sole ordinary refractive index (n_o). The sole n_o makes the OPL decreasing, and the ordinary defect modes shift to the shorter wavelength. However, the wavelength of defect modes in PBG remains unchanged, when the field-on and field-off are applied, because the polarization direction of the incoming light is in the y-axis and the same n_o contribution of the OPL is unchanged. We can conclude that the tunability of the defect modes is attributed to the change in refractive index (n_e or n_o) and its corresponding OPL.

In addition, the other nonchiral type is the twisted-nematic (TN) LC, in which the molecular orientation only exhibits 90° twist, acts like the optical polarization rotator, so that the incoming light passing through the TN LC is characterized by the rotation of polarization. The hybrid PC/TN LC structure was first demonstrated in 2010 [28]. The optical phenomena attributable to the PC/TN LC structure are quite different from the mechanism mentioned in the preceding paragraph. The 90° TN LC modes are divided into three groups based on the polarization angle β between the axis of the first polarizer and the director axis lying in the front substrate. They are classified as the ordinary-mode (O-mode), extraordinary-mode (E-mode), and mixed-mode (M-mode), and then TN satisfies the conditions of β = 90, 0, and 45°, respectively. The M-mode TN (abbreviated as MTN) combines both the polarization-rotation effect and birefringence effect. **Figure 3** shows the phenomenon of the wavelength shift of defect modes in two PC/TN cells impregnated with two different nematic LC materials. The defect modes of PC/TN for the ordinary ray are independent of the applied electric field. However, the wavelengths of defect modes in PBG with both E-mode and O-mode in PC/TN device are also shown as the blueshift, when we increase the applied voltage. This effect is unlike the ECB-based defect modes of PC/LC in the preceding paragraph. Compared with the O-mode and the E-mode in PC/TN, ECB-based PC exhibits more blueshift in wavelength because of the decrease in effective refractive index of defect layer significantly. This PC/TN result can be explained by Mauguin parameters [29]. A perfect adiabatic following in the TN LC makes the linearly polarized light to traverse the LC with the rotation of the LC molecular twist, makes the effective refractive index with the incident light nearly equal to n_e in E-mode and n_o in O-mode. The n_{eff} is no longer a constant in the O-mode TN LC cell, but becomes a weak dependence with applied voltage. Thus, the small shifts for the defect modes are demonstrated in the O-mode PC/TN cells. The most important is the integrated effect (M-mode) of defect modes in PC/TN device. The wavelengths of defect peaks in the M-mode are located at the same wavelength positions of the E-mode and O-mode because the defect peaks of the M-mode in PC/TN are contributed by two effects: the adiabatic following and birefringence effects of LC. Moreover, the intensity of the transmittance of

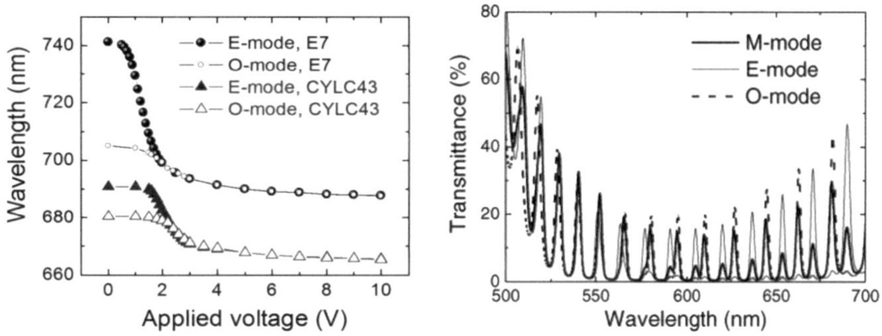

Figure 3.
The blueshift of defect modes for both E- and O-modes in PC/TN system, and transmittance of the PC/TN in the photonic bandgap with three different modes (adapted from [28]).

defect mode in E- or O-mode spreads to the other, making the intensity of the transmittance of the defect modes in the M-mode almost the same as both in E- or O-mode. Thus, we can observe that the M-mode is a superposition of both E-mode and O-mode. In addition, we observe carefully the spectra of E- and O-mode, small defect peaks accompanying the main defect modes are observed. This phenomenon is very different comparing with other types of PC/LC cells. Finally, no matter PC/LC or PC/TN, the PC-based nonchiral LC is the useful tool for optical devices or photonic applications.

2.2 Chiral-tilted homeotropic nematic liquid crystal-based photonic crystal devices

We introduce one of the chiral-type LC called bistable homeotropic nematic LC (BHN). Recently, the green energy concept is concerned with not only how to generate clean energy, but also how to save energy. Following this trend, PC devices with low energy consumption are highly desired. The novel device: a PC infiltrated with a BHN to achieve both the tunability of defect modes and the low energy consumption. The BHN bistable switching mechanisms involve the backflow and the frequency revertible dielectric anisotropy effect [30]. The PC/BHN can perform in two stable states, the tilted homeotropic (tH) and tilted twist (tT) states with nonvoltage. In addition, the two voltage-sustained states: the biased homeotropic (bH) and biased twist (bT) states at frequency 1 and 100 kHz, respectively, are proposed. **Figure 4** shows the LC configurations of the PC/BHN device in both tH state and tT state at 0 V; bH state and bT state at 10 V and 1 kHz. In addition, the switching between the bistable tH and tT states can be achieved by applying short voltage pulses to permit the BHN to pass through the intermediate states (bH and bT) [30]. In this BHN, the voltage-sustained states are necessary pathways for bistable operation served as the transient states. However, no voltage has to be applied to sustain the bistable tH and tT states, making PC device with green energy. The spectra of defect modes in the PC/BHN are interesting. The four states (tT, tH, bT, and bH) have different spectral profiles. The bH state at 10 V_{rms} exhibits the defect modes attributed to the ordinary refractive index, all the other states tT, tH, and bT have more peaks spectra caused by the effective refractive index. In addition, the intensity of the extraordinary defect modes in the tH state can be tuned by switching between the tH and bH states as intensity modulator. In the condition, the defect modes will diminish when the stable tH state transforms to the bH state at high voltages. This finding makes PC/BHN device with light-on and light-off states without any polarizers. In addition, **Figure 5** demonstrates the experimental spectra

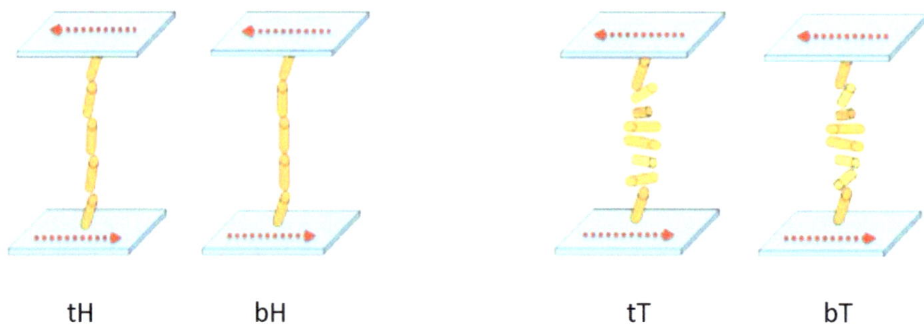

Figure 4.
LC configurations of the PC/BHN device. The tH state and tT state are at 0 V; bH and bT state are at 10 V and 1 kHz (adapted from [30]).

of a PC/BHN cell in the four bH, tH, bT, and tT states under the parallel polarizer. We can observe that the different spectra of bH, tH, bT, and tT are shown under the parallel polarizer.

Therefore, the PC/BHN device placed between a pair of linear polarizers has been proposed recently [30], and the axes of the two polarizers were parallel or perpendicular to each other in the new PC/BHN system. **Figure 5** demonstrates the bH state exhibited the defect modes corresponding to the sole ordinary refractive index n_o. From **Figure 5**, we can see that the defect modes of the bH state did not change with ϕ because of the LC molecules oriented vertically. In addition, the stable tT state was obtained from the bT state by turning off the high applied frequency. Comparing with the bT state, the tT state possessed a higher tilt angle, implying that the birefringence effect became more significant. Thus, both the ordinary and extraordinary components were conspicuous. Moreover, from **Figure 5**, the complementary in terms of defect mode wavelengths between the conditions of ϕ = 0 and 90° in the tT state was clearly shown. Furthermore, **Figure 6** shows the simulated spectra for the two tH and bH states under the parallel polarizer scheme with various polarization angles. We can observe that the differently distributed defect modes are shown. The calculation of the optical responses for both the bH and tH states by the transfer matrix method is also proposed [30]. The perfect agreement is satisfied between the simulated spectra and the experimental data (**Figure 6**). The profile width at half-maximum (FWHM) of the simulated defect mode peaks is narrower than the experimental one. This is attributable to minor experimental uncertainties like interface roughness and imperfect dielectric materials. Based on tunable optical properties of defect modes in PC/BHN, many photonic applications can be achieved.

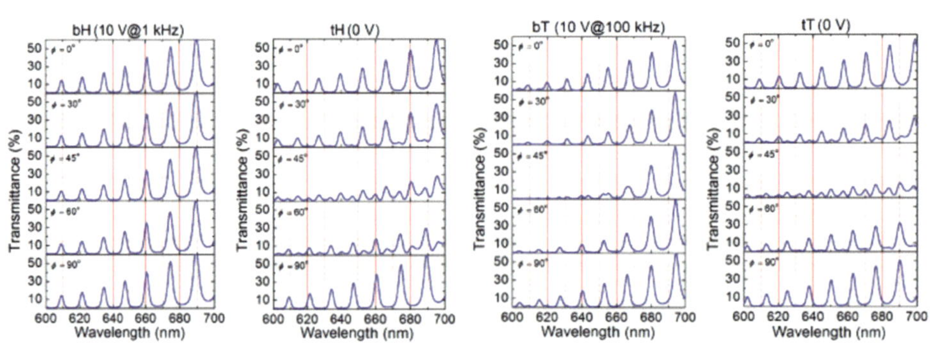

Figure 5.
Transmission spectra within the PBG of a PC/BHN device in four different states (bH, tH, bT, and tT) (adapted from [30]).

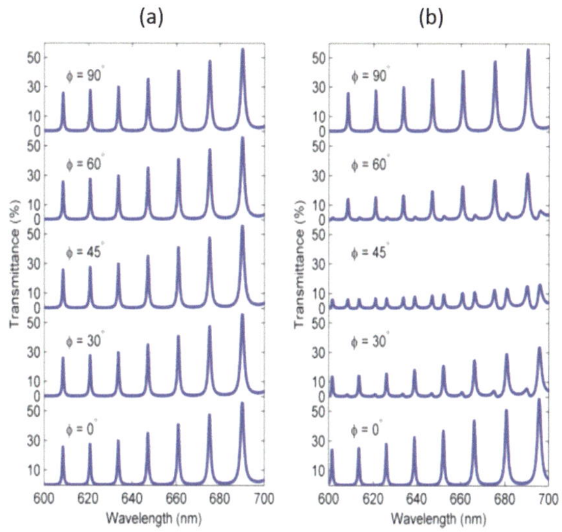

Figure 6.
Simulations of the transmission spectra of a PC/BHN device under the parallel polarizer at various polarization angles in (a) bH and (b) tH states (adapted from [30]).

In conclusion, a novel photonic structure PC/BHN with stopband, defect mode tunability and optical bistability have been shown. The BHN as a defect layer infiltrated within a PC was investigated recently. With the rich optical properties in the PC/BHN system and its association with the polarization effect, PC/BHN device further opens up new possible applications for the low-power consumption photonic devices in tunable spectral bandwidth and optical multichannel technologies.

2.3 Chiral nematic and cholesteric liquid crystal with photonic crystal devices

Comparing with the nematic LCs, the configuration of a cholesteric LC (CLC) or chiral LC with stacked layers shows a periodic helix of LC molecule. CLC characterized by a specific pitch length makes the structure regarded as 1D PC material by itself. The optical Bragg reflection or photonic band is the most important property in CLCs. Utilizing the special properties of periodic helix-induced photonic band in CLCs, many optic applications such as low-threshold single-mode laser with band edge excitation has been proposed. In addition, the special type of CLC is a dual frequency CLC (DFCLC). And DFCLC owns many special properties such as fast switching. Typical CLCs own bistable states, namely, the planar (P) and focal conic (FC) states. And then CLCs cannot directly switch from the FC state to the P state. Typically, this transition must be passing through an intermediate state: homeotropic (H) state or the transient P state [22–24]. However, the DFCLCs made of a DF nematic LC mixed with a chiral dopant could achieve fast and direct FC-to-P switching (~10 ms) [30]. In DFCLC, the dielectric anisotropy is positive and the LC director tends to be paralleled to the electrical field direction (tend to H state) when applied frequency below the crossover frequency. In contrast, the dielectric anisotropy is negative when the applying frequency is higher than the crossover frequency. And the DFLC director tends to be vertical to the field direction (tend to FC and P states). Therefore, we can use frequency-modulated voltage to switch between the bistable P and FC states reversibly, making the PC/CLC device with more tunable and switchable properties. Based on PC/CLC device, many applications such as intensity tunable and fast switching in the defect mode PC device. The detail structure of the PC/CLC device has been depicted in **Figure 7**. In addition,

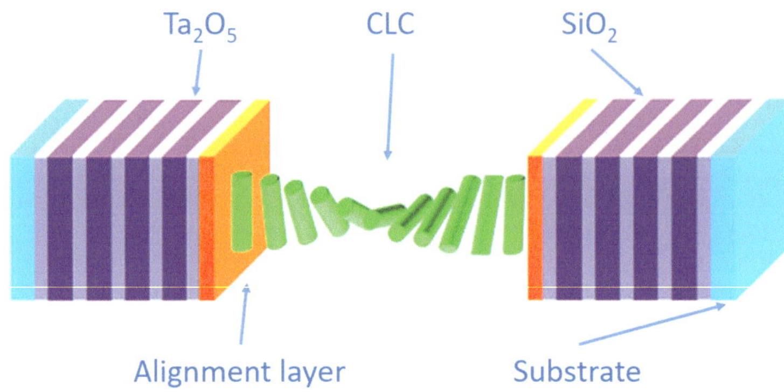

Figure 7.
The sandwich structure of the 1D PC/DFCLC device (adapted from [22–24]).

Figure 8 shows the transmission spectra of the PC/DFCLC device in three distinctive states (P, FC, and H states) at a various voltages. Among the three states, the P and FC states are optical stable states except the H state. Moreover, the stable P state can be achieved from the unstable H state by fast turning off the applied voltage or from the stable FC state by applying high frequency pulse [22–24]. In addition, **Figure 8** also shows that the hybrid PC/CLC device in the P state, which demonstrates a number of defect modes. Furthermore, the FC state of the hybrid PC device is exhibited when we apply voltage pulse of 20 V_{rms}. The optical intensity of the defect modes is very low in the FC state, and the spectra of defect modes in FC are shown in **Figure 8**. The light scattering properties of FC state make all defect modes turn off. This optical effect has the potential to expand as a fast switching light shutter application. Furthermore, the PC/CLC device will be in the H state when the voltage increases to 35 V_{rms}. And the most intense defect modes of H state are generated. **Figure 8** also shows the comparison of the spectra of defect modes between the P and H states in the PC/CLC device. We can observe that the blueshift of the defect modes of H state is shown and caused by the reduced effective index of refraction in the PC defect layer. It is interesting to observe the special phenomenon "complementary" in wavelengths of defect modes. This property can make the PC/CLC device as a tunable shutter in specific wavelengths of defect modes.

The interesting optical characteristics of PC/CLC devices have been investigated. By using the electrically controllable DFCLC materials as defect layer in the

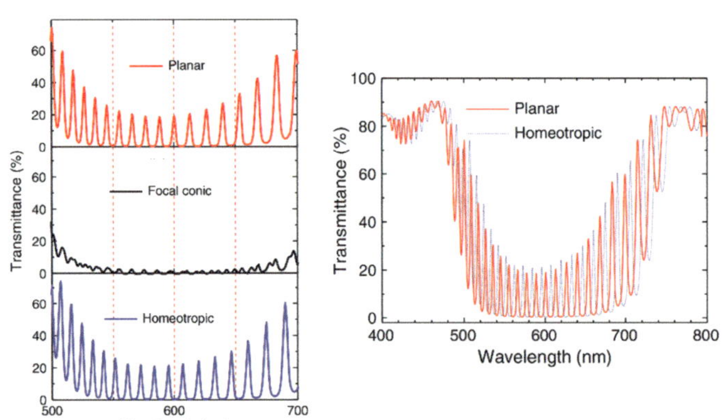

Figure 8.
Spectra of the PC/DFCLC device in the photonic bandgap in P, FC, and H states. The PC/DFCLC device is driven by various voltages. In addition, the PC/DFCLC in the photonic bandgap with two different sets of defect modes in both P and H states (adapted from [22–24]).

PC structure, PC/CLC device owns more powerful properties. Based on the three distinctive states of the CLC defect layer (stable P and FC states, and the voltage-sustained H state), the PC/CLC device exhibits different spectra in different LC states. In addition, the strength and wavelengths of the defect modes can be tuned by applying voltage and frequency. Moreover, the novel PC/CLC device is characterized by its fast switching between P and FC states. In the past research, the FC-to-P transition time is as short as 10 ms [22–24]. The wavelength and intensity tunability in the defect modes are more obvious comparing with other PC device. In addition, it requires no polarizers and is of low-power consumption because of bistability in P and FC states. This PC/CLC device is useful tool for photonic applications such as filter, light shutter, and optical modulator.

2.4 Tristable photonic crystal devices with polymer-stabilized cholesteric textures

In comparison with the typical CLC materials, with inclusion of a photo-polymerizable monomer into CLCs, which make CLC more powerful. The CLC/monomer composites own polymer networks to stabilize the CLC molecule, and we call the composite material as polymer-stabilized cholesteric texture (PSCT). The PSCTs can be employed in green energy devices due to the new stable state in the polymer-stabilized H state. This allows the bistable switching between the FC and P states in CLC become tristable P, FC, and H states potentially [22–24]. In the past, bistable PSCT shutters can also be switched between the H and FC states [31]. However, PSCTs are possible to own more than two stable modes. Recently, Hsiao et al. proposed the first tristable PSCT as a new PC device. **Figure 9** shows three photographs of P, FC, and H states and the corresponding micrographs of the PC/PSCT devices. In addition, the PC/PSCT is placed between two crossed polarizers in the tristable P, FC, and H states. We can discover that the colors are distinctive in the three different stable states. Firstly, the P state shows that the purple color due to the transmittance of defect modes are higher in red wavelength range. In addition, the light scattering FC state shows the multidomains of the PSCT and is presented in **Figure 9**. Moreover, the stable H state with the light leakage under crossed polarizers is also demonstrated in **Figure 9**. In addition, **Figure 10** demonstrates the spectra of defect modes in PC/PSCT device in three distinctive states (P, FC, and H states) at null voltage. The number of defect modes will increase with the increasing defect layer thickness [22–24]. Haiso et al. apply a fixed voltage (50 V_{rms}) at various frequencies to show the tristable states in PC/PSCT. From **Figure 10a**, we can observe the most

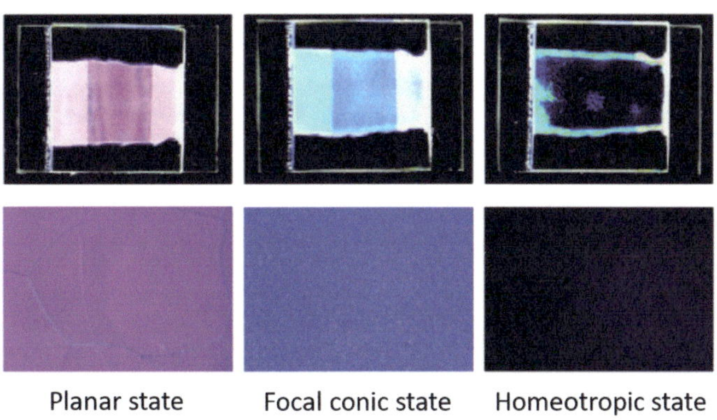

Figure 9.
Photographs and micrographs of the PC/PSCT device placed between crossed polarizers in P, FC, and H states at zero voltage (adapted from [22–24]).

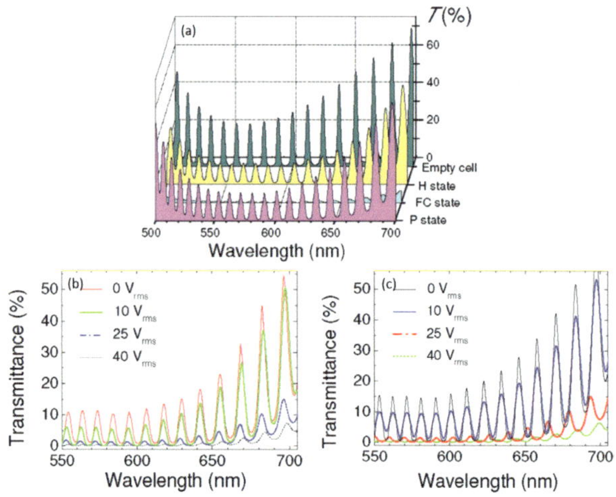

Figure 10.
(a) Transmission spectra of the empty PC cell and the PC/PSCT structure in three different states P, FC, and H states. In addition, (b) transmittance of the defect modes from H to FC in the PBG induced by a 100-kHz various voltage amplitudes. (c) Transmittance of the defect modes from P to FC in PC/PSCT induced by various voltage at a fixed frequency of 1 kHz (adapted from [22–24]).

intense defect modes in the empty PC cell because of the transparent air defect. With a PSCT defect layer embedded in the PC device, the PC/PSCT device initially in the H state and the more spectral defect windows in the PBG due to the higher (ordinary) refractive index n_o in the LC layer. And the FC state is demonstrated when a 30-kHz voltage pulse is applied. The lower transmission of the defect modes in the PBG is also shown in **Figure 10a**. We can employ the defect modes of FC to switch off the PC device by the light scattering property. When the frequency still increases to 100 kHz, the PC/PSCT will be in the P state. The redshifted defect modes is shown and the increasing defect mode number is exhibited (**Figure 10a**). Moreover, **Figure 10b** illustrates the spectra of the PC/PSCT device in the H and FC states induced by various voltage amplitudes at a fixed high frequency of 100 kHz. We can observe that the H state of the cell is the initial state. The intensity strength of the defect modes can be tuned by increasing the voltage. **Figure 10c** demonstrates the transmission spectra of the defect modes by applying various voltage of 0, 10, 25, 40 V_{rms} at a low frequency of 1 kHz. We can easily modulate the strength of defect modes between FC and P states. This powerful photonic device has the potential to expand optics applications, making it use as an electrically tunable device and optically tristable filter based on these special properties.

To conclude, the electrically tunable PC/PSCT devices have been investigated. In addition, the tunability is caused by the incorporation of a PSCT material as a new defect layer in PC structure. This hybrid PC/PSCT owns three stable P, FC, and H states. The electrically tunable PC device has been investigated, and it can be directly switched from one to another stable state by just applying a voltage pulse. Due to the tristability, the optical defect modes of PC/PSCT remain at zero voltages. This PC/PSCT composite device exhibits many different defect mode transmission spectra when we switch among P, FC, and H states. In addition, the intensity of the defect modes can be tuned by the amplitude of voltage as well as the wavelengths can be switched by the frequency in the H and P states. Based on the properties of tristable switching, wavelength controlling, and intensity tunability in the defect modes, the novel PC/PSCT device can be used as a low-power consumption optical filter, light shutter or an electrically intensity modulator without any polarizers, which let the PC/PSCT device more potential for applications.

3. Applications in liquid crystal-based photonic crystals

3.1 Electrically switchable liquid crystal-based photonic crystals for a white light laser

Laser source is the most unique light source with many special optical properties such as coherence and collimation. The laser emission needs both the elements: stimulating source and the gain media. Today, various solid and gas materials have been employed as gain media for lasing. However, white light lasers that span the visible spectrum (red, green, and blue colors) are important for lighting, imaging, and communication applications. Recently, the organic white light laser source was successfully demonstrated [32]. Recently, an inorganic semiconductor laser source has also been proposed with a monolithic multi-segment semiconductor nanostructure [33]. Huang et al. also shows that PC/CLC hybrid structure (**Figure 11a**) is a new way to achieve white light laser [34]. In addition, the complex stacking PC/CLC structure is designed (see **Figure 11a**) and can be simply coded as [GI(HL)4HH(LH)3]–P(D)P–[(HL)3HH(LH)4IG], where D means the dye-doped CLC (DDCLC); P is the polyimide alignment layer; H and L are the high and low refractive indices of dielectrics; G represents the glass substrate; and I is the ITO. In addition, the high and low refractive indices of dielectric materials are Ta_2O_5 (n_H = 2.18) and SiO_2 (n_L = 1.47). The configurations of the CLCs in three states (P, FC, and H states) are shown in **Figure 11b**. Note that the voltage V_1 leads to the FC state exhibiting an optical scattering property, and a larger voltage V_2 induces the H state. The transmission spectra of CLC and a PC substrate are also displayed in **Figure 12a**. In addition, the PBG is divided by a defect mode peak at the 640 nm of PBG. The Bragg reflection of DDCLC is located at right half of PBG in hybrid PC cell. The dye composition (C540A, PM580, and LD688) in the PC/DDCLC device was adjusted to fluoresce in three wavelengths (red, green, and blue lasing emissions). However, the artificial defect mode peak in the PC is at 446 nm, which allowed the pumping light to penetrate the PC cell. An organo-inorganic white light laser from PC/DDCLC composed of three colors red, green, and blue lasing emissions is therefore achieved, as displayed in **Figure 12b**. A genuine photo of the PC/DDCLC laser is shown in **Figure 12c**, which is accompanied by the CIE1931 chromaticity diagram. In addition, the color of red, green, and blue are mixed as the discrete

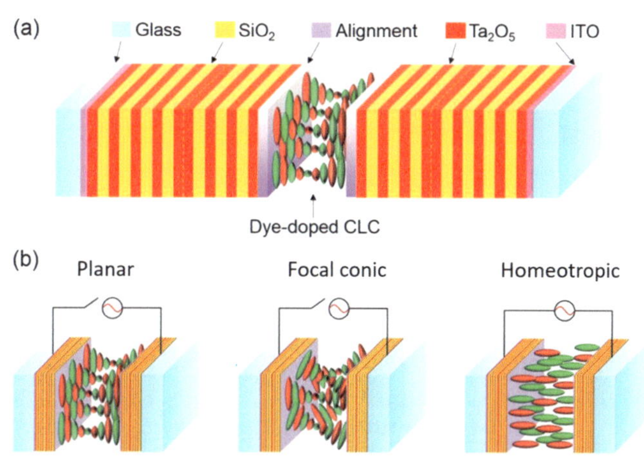

Figure 11.
Schematics of (a) the hybrid phonic structure and (b) the configurations of the three CLC states in the multilayers device (adapted from [32]).

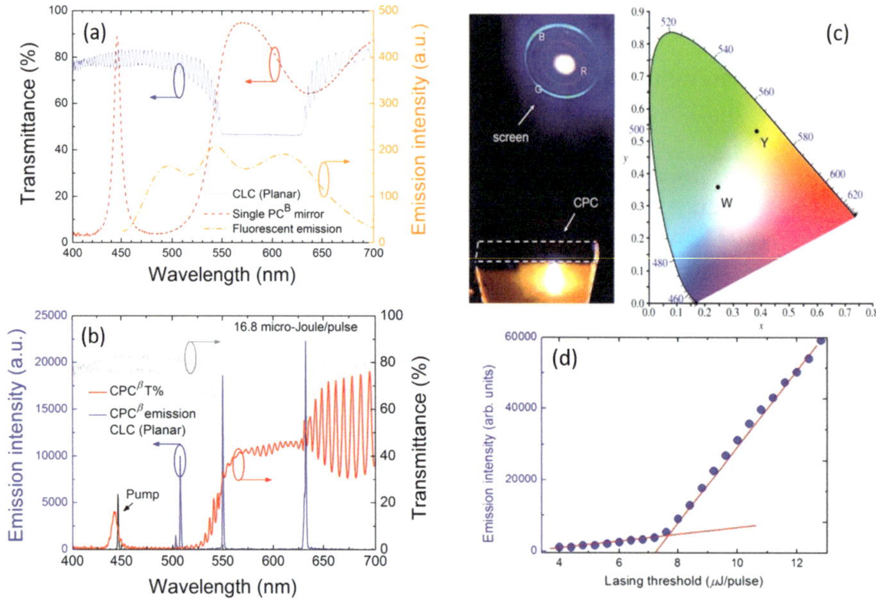

Figure 12.
Spectra of PC/DDCLC. (b) The white-light lasing spectrum and the spectra of PC and the CLC in the planar state. (c) Photograph of a tricolor laser device and the color space coordinates of the PC laser on the CIE 1931 chromaticity diagram. (d) the pumping energy-dependent the lasing emitted from PC/DDCLC device.

white light laser and depicted in **Figure 12c**. One can tell that the novel PC/DDCLC structure can be really lasing in white light. **Figure 12d** shows the relation between lasing intensity of PC device with the pumping energy. The threshold is about 7.4 μJ/pulse in this PC lasing device.

This is the first demonstration of a discrete white light source (three-colors: red, green, and blue) lasing. The organo-inorganic PC/DDCLC cannot only generate three colors in lasers with a single pump, but also be electrically switched among the three modes lasing. With such properties, lasing wavelength can be altered back and forth in a wavelength range and in a very short response time. In addition, PC/DDCLC lasing device is also cost effective, color tunable, and can be fabricated easily. Moreover, it has been shown that the PC device can be pumped using a simple CW laser. The ability to generate a single-color, two-color, three-color or white-light laser makes a new way to full color display, lighting, and other optics applications. By employing PC/DDCLC lasing device, a small size laser system can be achieved to make the proposed PC/DDCLC applications more feasible and potential.

3.2 Liquid crystal-based photonic crystals for pulse compression and signal enhancement in fluorescence applications

Multiphoton fluorescence microscopy, devised in 1990, has become an important technology for bio-applications. The improved axial depth and image penetration depth can reduce the bio-sample damage. This approach demonstrates potential applications in bio-imaging *in vivo* [35]. However, the high intensity of the excitation pulsed laser used in multiphoton fluorescence microscopy inducing the photo-damaging in specimens [36]. In the past, the method to reduce the photo damage in the bio-sample with strengthened multiphoton signal employs the excitation laser with narrower pulse widths. Recently, researchers found that the multiphoton fluorescence can be strengthened with the fluorescence signal, which is being proportional to the laser pulse widths. In 2001, both scientists McConnell and Riis [37] observed a seven-fold enhancement in two-photon fluorescence by the excitation-compressed laser pulses (~35 fs) [36]. However, by using this laser pulse compression techniques, a dispersion compensator is needed. Thus, the different microscope objectives contribute to different degrees of laser pulse broadening. Recently, Hsiao et al. propose the first PC device enabling on-specimen compression of excited laser pulse. The compression effect occurs after the laser light passing through the objective and photonic components. This will be significant to enhance the multiphoton fluorescence signal. In addition, the PC devices combining with LC materials as defect layer can make the device with the tunable property. From now on, the LC-based PCs for the pulse compression and signal enhancement in multiphoton fluorescence have been proposed.

Moreover, in order to measure the pulse widths through the PC device on-specimen, an optical autocorrelator was employed in the multiphoton fluorescence microscopy. This new approach allows us to detect the autocorrelation signal at the focal plane of the objective, which are shown in **Figure 13a**. In addition, the Ti:sapphire laser is sent through a 50% beam splitter, and one of the optical beams passes through a variable delay line system. Moreover, the multiphoton fluorescence signals can be detected by a photomultiplier tube (PMT). The autocorrelation signal traces with a peak-to-background ratio of the interferences are 8:1 and shown in **Figure 13b** and **c**. We can observe that the envelope of the interferences is fitted to the function of Gaussian. The original pulse width of the commercial Ti-Sapphire laser is about 100 fs. However, the laser pulse width was broadened to be 270 fs after passing through the optical components and objective, (**Figure 13b**). In addition, the laser pulse width decreases in a nonlinear fashion when we increase the power. The most important is the shortest pulse duration is 30 fs (**Figure 13d**). **Figure 14a** shows the images of the red channel at different exposure time under the operating power 40 mW with device and 150 mW without device. If the PC/LC device was not used, the photo damage becomes apparent when the exposure time beyond 1.5 h. If the proposed PC device was

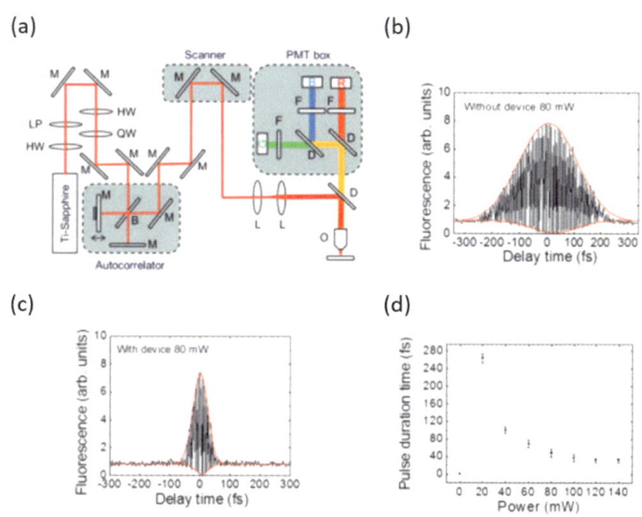

Figure 13.
(a) The system design of optical autocorrector in multiphoton fluorescence microscopy. HW is half-wave plates, LP stands for linear polarizers, QW is quarter-wave plates, M means mirrors, B is a beam splitter, L is lenses, O stands for the objective, D means dichroic mirrors, and F stands for filters; (b) and (c) are the autocorrection traces without and with the PC devices; (d) pulse duration time versus the applied laser powers (adapted from [38]).

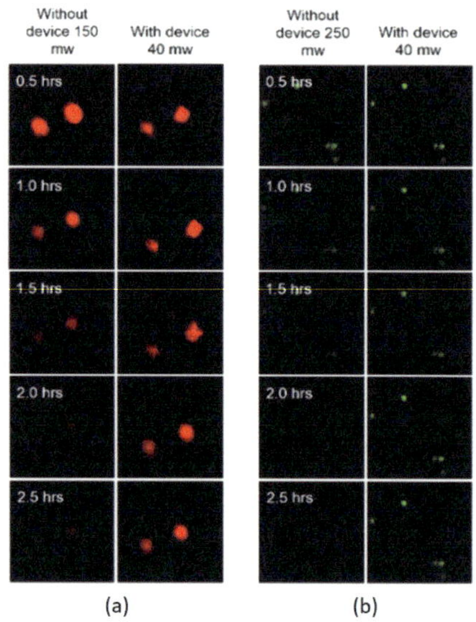

Figure 14.
The photo images of (a) the red and (b) the green fluorescent balls at different illumination time with and without the PC/LC device (adapted from [38]).

used, the photo damage effect can be easily reduce and lower the excitation power, which is applied to achieve the same signal intensity. We can observe that PC/LC device can efficiently reduce the both operation power and photo damage. In addition, **Figure 14b** shows the same effect of photo damage reducing in the green fluorescent balls under applied voltage 10 V. Thus, this novel PC/LC device is much more powerful for bio-imaging in photo damage reducing. In addition, this PC/LC device for laser pulse compression does not need any dispersion correction, making the biologists easy to use the PC/LC device.

In conclusion, Dr. Hsiao used a PC/LC to compress the laser pulse, exhibiting a 15-fold enhancement of the fluorescence. Without any dispersion compensator, the PC/LC device can be more convenient for nonphotonic researchers. By using the both pulse compression effect of PCs and the tunability of LCs, the PC/LC device shows a new way to enhance the multiphoton fluorescence microscopy with lower photo damage.

3.3 Photo-manipulated photonic devices based on tristable chiral-tilted homeotropic nematic liquid crystal

In recent years, energy saving materials have attracted much attention from scientists. The energy saving materials need to own excellent optical stability and do not require constantly applied energy. Based on the stable state natural properties, bistable LC devices are shown and can be used as e-books or e-papers. Recently, compared with the bistable in LC modes, tristable or multi-stable LCs have been scarcely proposed. Historically, the tristable LC mode was first exhibited in a ferroelectric LC system in 1988, and then the first tristable CLC device was later proposed by Hsiao et al. [22–24]. However, the stability of a LC state is very pressure-sensitive of the LCs to be bistable or tristable. Hsiao et al. demonstrate a new tristable optical composite—dye-doped tristable chiral-tilted homeotropic nematic (TCHN). This TCHN mode is extended from the technique of BHN mode. In comparison with BHN, TCHN adopts a common nematic LC material instead of DFLC material; it possesses an additional stable state and is stress-insensitive in stable states. Recently, Huang et al. proposed the spectral properties of an optically switchable TCHN incorporated as a tunable defect layer in PC structure. By controlling the polarization angle of the incident light as well as the intensity ratio between UV and green light, the tunable transmission characteristics of defect modes in the PC/TCHN were obtained. The hybrid PC structure realizes photo-tunability of defect mode peaks within the photonic bandgap. The PC/TCHN has much potential for many photonic applications such as a low-power consumption filter and an optically controllable intensity modulator device.

A schematic of the hybrid PC/TCHN structure is shown in **Figure 15a**. The optical switching of TCHN is among the tT, fingerprint (FP), and tH states as the new tristable PC device and is displayed in **Figure 15b**. The chiral bis(azobenzene) molecule is photoresponsive and is used in TCHN system. The chiral has two azo linkages to confer two distinct isomeric conformations: the rod like *trans* form and the bent *cis* form. The photo-induced unwinding effect caused by the *tran*-to-*cis* isomerization exposed to UV light and the winding effect due to the *cis*-to-*trans* isomerization under green light illumination. Based on the mechanism, we can optically switch TCHN among the tT, FP, and tH states. The experimental spectra of the PC/TCHN devices were measured under irradiation by controlling both green light and UV light. In Hsiao's paper, the used green light is from LED at wavelength of 524 nm and the UV light is from a UV LED at wavelength of 365 nm. Based on the two mixed irradiation lights, the PC/TCHN can be switched among three stable states (tT, FP, and tH). In addition, **Figure 15** shows the micrographic optical textures of the state-transform process among tristable tT, FP, and tH states under a crossed-polarizing microscope. Moreover, **Figure 15a, c**, and **e** shows the micrographic optical textures of tT, FP, and tH states, respectively. The tH state is completely dark state in the crossed-polarizer scheme because of no birefringence effect. The tT state is a bright optical texture with some rubbing traces, and the FP texture involves the lying helix structure of chiral nematic molecules as shown. The transmission spectra of defect modes within the PBG at various irradiations are also proposed. **Figure 16** demonstrates the transmittance spectra of defect modes that are controlled by the UV light with various intensities. In addition, the intensity of green light is fixed at 2.02 mW/cm^2. From **Figure 16f**, we can observe that the PC/TCHN device is exhibited in the tT state when intensity of UV is 0 mW/cm^2 because of the azo chiral dopant being in the trans-form. However, the cis-form dopant molecules will increase when intensity of UV is strengthened gradually. Because the high-pretilt angle of LC and particular d/p conditions,

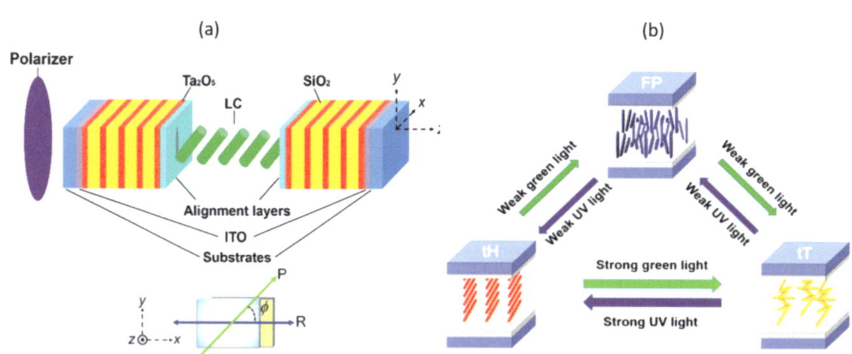

Figure 15.
(a) Sandwich structure of the PC/TCHN device. The arrows in the device's front view show the transmission axis of the polarizer (P) and rubbing direction (R). (b) Operating mechanisms of photo-induced TCHN switching among the tH, FP and tT states (adapted from [34]).

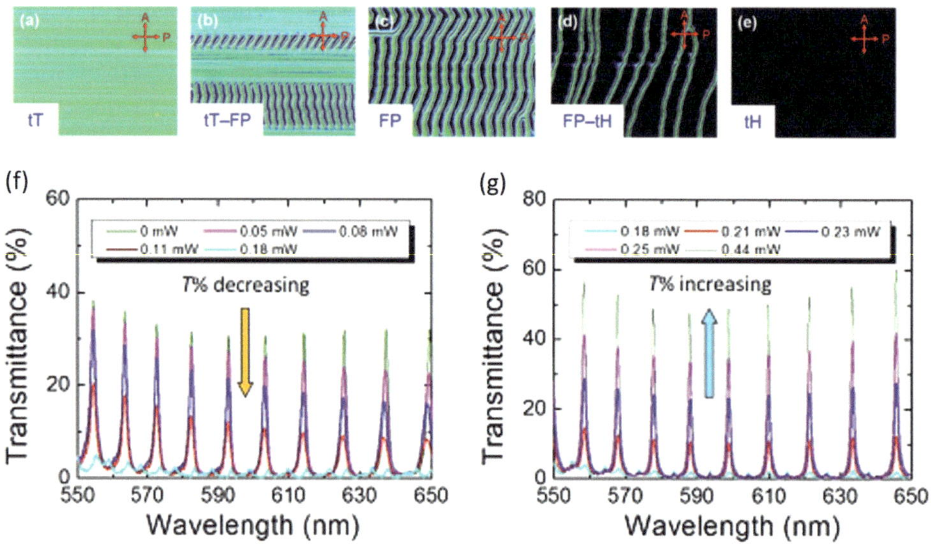

Figure 16.
The transition process among three stable states by increasing irradiation ratios between UV and green light. (a) The tT state; (b) the coexistence of tT and FP states; (c) the FP state; (d) the coexistence of FP and tH states; (e) the pure tH state of a TCHN. The arrows indicate the transmission axes of the polarizer (P) and analyzer (A). In addition, photo-manipulated transmittance of the defect modes under irradiation by various UV powers. (f) From the tT to the FP state and (g) from the FP to tH state (adapted from [34]).

the FP texture will appear. Then, we can see that the intensities of defect mode vanish because of strong scattering. However, the defect mode shifts to distinct wavelengths and emerges again with enhanced intensity of UV. When the azo chiral molecules become fully converted to the trans-form at 0.44 mW/cm^2, the PC/TCHN comes to the tH state and the transmittance of defect modes reaches the highest intensity as shown in **Figure 16g**. Based on these optical properties, the PC/TCHN device can be used as tunable optical filters.

In conclusion, a novel concept of photo-switchable PC/TCHN devices is proposed. In comparison with its bistable counterpart BHN, the TCHN can be prepared using a regular nematic host LC material (e.g., E7 in experimental from Huang et al.). The PC/TCHN device owns optical tristability and tunability in wavelength and intensities of the defect modes by photo manipulating. By adjusting the ratio of the UV and green light intensity, the defect modes of PC/TCHN not only show a variation in spectral amplitude through the stable FP state, but also can control the wavelengths between the stable tT and tH states. This novel PC/TCHN mode and the special properties of TCHN material should be fully developed for next potential photonic applications [39–72].

4. Conclusions

In this chapter, PC/NLC and PC/chiral LC devices, which exhibit special photonic applications and several fascinating features have been reviewed. In addition, the optical properties of defect modes of PC switching among each state are reported. In Section 2.1, "Photonic crystals with a nonchiral nematic liquid crystal," the mechanism of defect modes in nonchiral PC is well-described by OPL. The shifting can be understood from a change in n_{eff} under applied voltage. Section 2.2 "Chiral-tilted homeotropic nematic liquid crystal-based photonic crystal devices," propose the new type of chiral nematic LC (BHN) in PC device. The optical properties and switching mechanism of four tH, bH, bT, and tT states of the PC/BHN device are proposed. In addition, Section 2.3 "Chiral nematic and cholesteric liquid crystal with photonic crystal devices," and Section 2.4 "Tristable photonic-crystal devices with polymer-stabilized cholesteric textures." are the typical chiral nematic or CLC structure within PC devices. The three states of P, FC, and H states from CLCs are used for defect modes controlling in PCs. Thus, the configuration of many types of PC/CLC cell applications, including the design of PC multilayers for applications is schematically depicted. I introduce the three PC applications; Session 3.1 introduces "Electrically switchable liquid crystal-based photonic crystals for a white-light laser." The first PC/LC white light laser is invented, and the mechanism of the laser chip is shown. Session 3.2 shows "Liquid crystal-based photonic crystals for pulse compression and signal enhancement in fluorescence applications," and it is the first PC/LC device to achieve the pulse compression. From now on, the PC/LC can be as fluorescence enhancement applications. Session 3.3 demonstrates "Photo-manipulated photonic devices based on tristable chiral-tilted homeotropic nematic liquid crystal." In addition to this, many PC/LC applications, such as optical devices, tunable optical filters, tunable optical modulators, and so on, are also developed in recent year. In conclusion, PC/LC devices are new tools to understand and modulate light. It holds good potential to become a useful tool in our daily life.

Acknowledgements

This work was financially supported by the Ministry of Science and Technology, Taiwan, under grant No. 107-2218-E-038-007-MY, and Taipei Medical University, Taiwan, under grant No. TMU106-AE1-B49.

Author details

Yu-Cheng Hsiao
Graduate Institute of Biomedical Optomechatronics, College of Biomedical Engineering, Taipei Medical University, Taipei, Taiwan

*Address all correspondence to: ychsiao@tmu.edu.tw

References

[1] John S. Strong localization of photons in certain disordered dielectric superlattices. Physical Review Letters. 1987;**58**(23):2486-2489

[2] Yablonovitch E. Inhibited spontaneous emission in solid-state physics and electronics. Physical Review Letters. 1987;**58**(20):2059-2062

[3] Fleming JG, Lin S-Y. Three-dimensional photonic crystal with a stop band from 1.35 to 1.95 μm. Optics Express. 1999;**24**(1):49-51

[4] Imada M, Noda S, Chutinan A, Tokuda T, Murata M, Sasaki G. Coherent two-dimensional lasing action in surface-emitting laser with triangular-lattice photonic crystal structure. Applied Physics Letters. 1999;**75**(3):316-318

[5] Knight JC. Photonic crystal fibres. Nature. 2003;**424**(6950):2486-2589

[6] Nelson BE, Gerken M, Miller DAB, Piestun R, Lin C-C, Harris JS. Use of a dielectric stack as a one-dimensional photonic crystal for wavelength demultiplexing by beam shifting. Optics Express. 2000;**25**(20):1502-1504

[7] Park SH, Xia B, Gates Y. A three-dimensional photonic crystal operating in the visible region. Advanced Materials. 1999;**11**(6):462-466

[8] Painter O, Lee RK, Scherer AY, O'Brien JDA, Dapkus PD, Kim I. Two-dimensional photonic band-gap defect mode laser. Science. 1999;**284**(5421):1819-1821

[9] Knight JC, Broeng J, Birks TA, Russell PSJ. Photonic band gap guidance in optical fibers. Science. 1998;**282**(5393):1476-1478

[10] Blanco A, Chomski E, Grabtchak S, Ibisate M, John S, Leonard SW, et al. Large-scale synthesis of a silicon photonic crystal with a complete three-dimensional bandgap near 1.5 micrometers. Nature. 2000;**405**(6785):437-440

[11] Chow E, Lin SY, Johnson SG, Villeneuve PR, Joannopoulos JD, Wendt JR, et al. Three-dimensional control of light in a two-dimensional photonic crystal slab. Nature. 2000;**407**(6807):983-986

[12] Ozaki R, Matsui T, Ozaki M, Yoshino K. Electro-tunable defect mode in one-dimensional periodic structure containing nematic liquid crystal as a defect layer. Japanese Journal of Applied Physics. 2002;**41**(12B):L1482-L1484

[13] Matsuhisa Y, Ozaki R, Takao Y, Ozaki M. Linearly polarized lasing in one dimensional hybrid photonic crystal containing cholesteric liquid crystal. Journal of Applied Physics. 2007;**101**(3):033120

[14] Matsuhisa Y, Ozaki R, Yoshino K, Ozaki M. High Q defect mode and laser action in one-dimensional hybrid photonic crystal containing cholesteric liquid crystal. Applied Physics Letters. 2006;**89**(10):101109

[15] Ozaki R, Matsuhisa Y, Ozaki M, Yoshino K. Low driving voltage tunable laser based on one-dimensional photonic crystal containing liquid crystal defect layer. Molecular Crystals and Liquid Crystals. 2005;**441**(1):87-95

[16] Ozaki R, Ozaki M, Yoshino K. Defect mode switching in one-dimensional photonic crystal with nematic liquid crystal as defect layer. Japanese Journal of Applied Physics. 2003;**42**(6B):L669-L671

[17] Park B, Kim M, Kim SW, Kim IT. Circularly polarized unidirectional

lasing from a cholesteric liquid crystal layer on a 1-D photonic crystal substrate. Optics Express. 2009;**17**(15):12323-12331

[18] Arkhipkin VG, Gunyakov VA, Myslivets SA, Zyryanov VY, Shabanov VF. Angular tuning of defect modes spectrum in the one-dimensional photonic crystal with liquid-crystal layer. European Physical Journal E. 2007;**24**(3):297-302

[19] Arkhipkin VG, Gunyakov VA, Myslivets SA, Gerasimov VP, Zyryanov VY, Vetrov SY, et al. One-dimensional photonic crystals with a planar oriented nematic layer: Temperature and angular dependence of the spectra of defect modes. Journal of Experimental and Theoretical Physics. 2008;**106**(2):388-398

[20] Zyryanov VY, Gunyakov VA, Myslivets SA, Arkhipkin VG, Shabano VF. Electrooptical switching in a one-dimensional photonic crystal. Molecular Crystals and Liquid Crystals. 2008;**488**(1):118-126

[21] Zyryanov VY, Myslivets SA, Gunyakov VA, Parshin AM, Arkhipkin VG, Shabanov VF, et al. Magnetic-field tunable defect modes in a photonic-crystal/liquid-crystal cell. Optics Express. 2010;**18**(2):1283-1288

[22] Hsiao Y-C, Hou C-T, Zyryanov VY, Lee W. Multichannel photonic devices based on tristable polymer-stabilized cholesteric textures. Optics Express. 2011;**19**(8):7349-7355

[23] Hsiao Y-C, Tang C-Y, Lee W. Fast-switching bistable cholesteric intensity modulator. Optics Express. 2011;**19**(10):9744-9749

[24] Hsiao Y-C, Wu C-Y, Chen C-H, Zyryanov VY, Lee W. Electro-optical device based on photonic structure with a dual-frequency cholesteric liquid crystal. Optics Letters. 2011;**36**(14):2632-2634

[25] Wang CT, Chen CW, Yang TH, Nys I, Li CC, Lin TH, et al. Electrically assisted bandedge mode selection of photonic crystal lasing in chiral nematic liquid crystals. Applied Physics Letters. 2018;**112**(4):043301

[26] Ye L, Wang Y, Feng Y, Liu B, Gu B, Cui Y, et al. Thermally switchable photonic band-edge to random laser emission in dye-doped cholesteric liquid crystals. Laser Physics Letters. 2018;**15**(3):035002

[27] Ozaki R, Ozaki M, Yoshino K. Defect mode in one-dimensional photonic crystal with in-plane switchable nematic liquid crystal defect layer. Japanese Journal of Applied Physics. 2004;**43**(11B):L1477-L1479

[28] Lin Y-T, Chang W-Y, Wu C-Y, Zyryanov VY, Lee W. Optical properties of one-dimensional photonic crystal with a twisted-nematic defect layer. Optics Express. 2010;**18**(26):26959-26964

[29] Gooch CH, Tarry HA. The optical properties of twisted nematic liquid crystal structures with twist angles < 90 degrees. Journal of Physics D: Applied Physics. 1975;**8**(13):1575-1584

[30] Hsiao YC, Zou YH, Timofeev IV, Zyryanov VY, Lee W. Spectral modulation of a bistable liquid-crystal photonic structure by the polarization effect. Optical Materials Express. 2013;**3**(6):821-828

[31] Ma J, Shi L, Yang D-K. Bistable polymer stabilized cholesteric texture light shutter. Applied Physics Express. 2010;**3**(2):021702

[32] Ha NY et al. Simultaneous red, green, and blue lasing emissions in a single pitched cholesteric liquid

crystal system. Advanced Materials. 2008;**20**:2503-2507

[33] Fan F, Turkdogan S, Liu Z, Shelhammer D, Ning CZ. A monolithic white laser. Nature Nanotechnology. 2015;**10**:796-803

[34] Huang KC, Hsiao YC, Timofeev IV, Zyryanov VY, Lee W. Photo-manipulated photonic bandgap devices based on optically tristable chiral-tilted homeotropic nematic liquid crystal. Optics Express. 2016;**24**(22):25019-25025

[35] Helmchen F, Denk W. Deep tissue two-photon microscopy. Nature Methods. 2005;**2**(12):932-940

[36] Hopt A, Neher E. Highly nonlinear photodamage in two-photon fluorescence microscopy. Biophysical Journal. 2001;**80**(4):2029-2036

[37] McConnell G, Riis E. Two-photon laser scanning fluorescence microscopy using photonic crystal fiber. Journal of Biomedical Optics. 2004;**9**(5):922-928

[38] Hsiao YC. Liquid crystal-based tunable photonic crystals for pulse compression and signal enhancement in multiphoton fluorescence. Optical Materials Express. 2016;**6**(6):1929-1934

[39] Bao R, Liu C-M, Yang D-K. Smart bistable polymer stabilized cholesteric texture light shutter. Applied Physics Express. 2009;**2**:112401

[40] Berreman DW, Heffner WR. New bistable cholesteric liquid-crystal display. Applied Physics Letters. 1980;**37**(1):109-111

[41] Bobrovsky AY, Boiko NI, Shibaev VP, Wendorff JH. Cholesteric mixtures with photochemically tunable circularly polarized fluorescence. Advanced Materials. 2003;**15**(4):282-287

[42] Choi SS, Morris SM, Huck WTS, Coles HJ. Electrically tuneable liquid crystal photonic bandgaps. Advanced Materials. 2009;**21**(38):3915-3918

[43] Hsiao YC, Wang HT, Lee W. Thermodielectric generation of defect modes in a photonic liquid crystal. Optics Express. 2014;**22**(3):3593-3599

[44] Hsiao YC, Lee W. Lower operation voltage in dual-frequency cholesteric liquid crystals based on the thermodielectric effect. Optics Express. 2013;**21**(20):23927-23933

[45] Hsiao YC, Sung YC, Lee MJ, Lee W. Highly sensitive color-indicating and quantitative biosensor based on cholesteric liquid crystal. Biomedical Optics Express. 2015;**6**(12):5033-5038

[46] Hsiao YC, Timofeev IV, Zyryanov VY, Lee W. Hybrid anchoring for a color-reflective dual-frequency cholesteric liquid crystal device switched by low voltages. Optical Materials Express. 2015;**5**(11):2715-2720

[47] Hsiao YC, Huang KC, Lee W. Photo-switchable chiral liquid crystal with optical tristability enabled by a photoresponsive azo-chiral dopant. Optics Express. 2017;**25**(3):2687-2693

[48] Hsiao YC, Lee W. Polymer stabilization of electrohydrodynamic instability in non-iridescent cholesteric thin films. Optics Express. 2015;**23**(17):22636-22642

[49] Hsiao YC, Yang ZH, Shen D, Lee W. Red, green, and blue reflections enabled in an electrically tunable helical superstructure. Advanced Optical Materials. 2018;**6**(5):1701128

[50] Hsu J-S. Stability of bistable chiral-tilted homeotropic nematic liquid crystal displays. Japanese Journal of Applied Physics. 2007;**46**(11):7378-7381

[51] Hsu J-S, Liang B-J, Chen S-H. Bistable chiral tilted-homeotropic nematic liquid crystal cells. Applied Physics Letters. 2004;**85**(23):5511-5513

[52] Hsu J-S, Liang B-J, Chen S-H. Dynamic behaviors of dual frequency liquid crystals in bistable chiral tilted-homeotropic nematic liquid crystal cell. Applied Physics Letters. 2007;**89**(5):091520

[53] Huang CY, Fu K-Y, Lo K-Y, Tsai M-S. Bistable transflective cholesteric light shutters. Optics Express. 2003;**11**(6):560-565

[54] Huang Y, Zhou Y, Doyle C, Wu S-T. Tuning the photonic band gap in cholesteric liquid crystals by temperature-dependent dopant solubility. Optics Express. 2006;**14**(3):1236-1244

[55] Jhun CG, Chen CP, Lee UJ, Lee SR, Yoon TH, Kim JC. Tristable liquid crystal display with memory and dynamic operating modes. Applied Physics Letters. 2006;**89**(12):123507

[56] Kasano M, Ozaki M, Yoshino K, Ganzke D, Haase W. Electrically tunable waveguide laser based on ferroelectric liquid crystal. Applied Physics Letters. 2003;**82**(23):4026-4028

[57] Krauss TF, La Rue D, Rchard M, Brand S. Two-dimensional photonic bandgap structures operating at near-infrared wavelengths. Nature. 1996;**383**(6602):699-702

[58] Miroshnichenko AE, Brasselet E, Kivshar YS. All-optical switching and multistability in photonic structures with liquid crystal defects. Applied Physics Letters. 2008;**92**(25):253306

[59] Miroshnichenko AE, Brasselet E, Kivshar YS. Light-induced orientational effects in periodic photonic structures with pure and dye-doped nematic liquid crystal defects. Physical Review A. 2008;**78**(5):053823

[60] Morris SM, Ford AD, Pivnenko MN, Coles HJ. Enhanced emission from liquid-crystal lasers. Journal of Applied Physics. 2005;**97**(2):023103

[61] Ozaki R, Moritake H, Yoshino K, Ozaki M. Analysis of defect mode switching response in one-dimensional photonic crystal with a nematic liquid crystal defect layer. Journal of Applied Physics. 2007;**101**(3):033503

[62] Timofeev IV, Lin Y-T, Gunyakov VA, Myslivets SA, Arkhipkin VG, Ya VS, et al. Voltage-induced defect mode coupling in a one-dimensional photonic crystal with a twisted-nematic defect layer. Physical Review E. 2012;**85**(1):011705

[63] Wen C-H, Wu S-T. Dielectric heating effects of dual-frequency liquid crystals. Applied Physics Letters. 2005;**86**(23):231104

[64] Wu S-T, Wu C-S. Mixed-mode twisted nematic liquid crystal cells for reflective displays. Applied Physics Letters. 1996;**68**(11):1455-1457

[65] Wu C-Y, Zou Y-H, Timofeev I, Lin Y-T, Zyryanov VY, Hsu J-S, et al. Tunable bi-functional photonic device based on one-dimensional photonic crystal infiltrated with a bistable liquid-crystal layer. Optics Express. 2011;**19**(8):7349-7355

[66] Xu M, Yang D-K. Dual frequency cholesteric light shutters. Applied Physics Letters. 1997;**70**(6):720-722

[67] Yao I-A, Yang C-L, Chen C-J, Pang J-P, Liao S-F, Li J-H, et al. Bistability of splay and π-twist states in a chiral-doped dual frequency liquid crystal cell. Applied Physics Letters. 2009;**94**(7):071104

[68] Yeh PH, Gu C. Normal modes of propagation in a general TN-LC. In: Optics of Liquid Crystal Display. Canada: John Wiley & Sons. ISBN 0-471-18201-X; 1999. pp. 130-136

[69] Yeh PH, Gu C. Light propagation in uniaxial media. In: Optics of Liquid Crystal Display. Canada: John Wiley & Sons. ISBN 0-471-18201-X; 1999. pp. 63-68

[70] Yeung Fion S-Y, Xie F-C, Wan Jones T-K, Lee FK, Tsui Ophelia KC, Sheng P, et al. Liquid crystal pretilt angle control using nanotextured surfaces. Journal of Applied Physics. 2006;**99**(12):124506

[71] Yin Y, Shiyanovskii SV, Lavrentovich OD. Electric heating effects in nematic liquid crystals. Journal of Applied Physics. 2006;**100**(2):024906

[72] Yokoyama S, Mashiko S, Kikuchi H, Uchida K, Nagamura T. Laser emission from a polymer-stabilized liquid-crystalline blue phase. Advanced Materials. 2006;**18**(1):48-51

Chapter 5

The Mid-Infrared Photonic Crystals for Gas Sensing Applications

Tahere Hemati and Binbin Weng

Abstract

Mid-infrared spectrum is known as the "molecular fingerprint" region, where most of the trace gases have their identical absorption patterns. Photonic crystals allow the control of light-matter interactions within micro/nanoscales, offering unique advantages for gas analyzing applications. Therefore, investigating mid-infrared photonic crystal based gas sensing methods is of significant importance for the gas sensing systems with high sensitivity and portable footprint features. In recent various photonic crystal gas sensing techniques have been developing rapidly in the mid-infrared region. They operate either by detecting the optical spectrum behavior or by measuring the material properties, such as the gas absorption patterns, the refractive index, as well as the electrical conductivities. Here, we will brief the progress, and review the above-listed photonic crystal approaches in the mid-infrared range. Their uniqueness and weakness will both be presented. Although the technical level for them has not been ready for commercialization yet, their small size, weight, power consumption and cost (SWaP-C) features offer great values and indicate their enormous application potentials in future, especially under the stimulation of the newly emerging technology "Internet of Things" which heavily relies on modern SWaP-C sensor devices.

Keywords: mid-infrared, photonic crystal, gas sensing, non-linear light, nanophotonics

1. Introduction

Chemical sensing, especially the trace-gas detection, is of a significant importance for a wide range of applications in practical fields including the medical inspection, the environmental monitoring, the manufacturing control, as well as the nation security surveillance. For example, controlling respiratory gases in the medical sector can be a matter of life or death [1, 2]. Monitoring the toxic, explosive and air polluting gases such as NO_x, CO_2, CO, CH_4, and O_3 is vital to prevent harms to human communities [2]. Nowadays, there are many choices of gas sensors based on different methods such as semiconductor [3, 4], catalytic [5], field effect [6], electrochemical [7], and optical gas sensors [8, 9]. Among them, the optical-based gas sensors have been well-received as a fast, precise and reliable technique, which has a long life expectancy, immunes to all chemical poisoning,

IntechOpen

and requires low maintenance for high-precision operation [9]. Furthermore, with the rise of the emerging technology, Internet of Things (IoTs), such high standard of performance requirements become more and more desired from the industrial world [10]. It is also necessary to point out that, besides of the aforementioned features, the importance of reducing the size, weight, power, and cost (SWaP-C) to enable the chip scale integration is revealed clearly in the development of IoT technologies [11]. Unfortunately, limited by the Beer-Lambert Law, to achieve a suitable sensitivity, a large optical interaction length is required in the conventional optical gas sensor systems. This long path of interaction makes the optical gas sensors relatively large and costly to be manufactured [12], which consequently sets a fundamental barrier to satisfy the SWaP-C requirements. Therefore, in such a high-speed technology innovating era, more and more researchers have paid significant attention to develop new technologies that can overcome these limitations.

Photonic crystals (PhCs), which can be regarded as one of the most advanced modern photonic technologies [13–16] was firstly proposed by Yablonvitch and John since 1980s [17, 18]. Owing to the unique non-linear optical dispersive properties [19, 20], PhCs become a powerful tool for control and manipulation of light-matter interactions on micrometer length scales [21, 22]. Considering the aforementioned optical sensing limitations, this technology is capable of addressing the size issue, and potentially suitable for on-chip gas detection applications. In recent year, the research on PhC-based gas sensing research has developed rapidly. Various sensing techniques have been proposed by using PhCs to detect chemicals in gas, vapor and even liquid environment [23–25]. Several review articles have been reported as well [26, 27]. For example, in 2013, Xu et al. [27] investigated different photonic crystal structures, such as morpho-butterfly wing, porous silicon PhCs, multilayer PhC films, colloidal PhCs, and Inverse opal colloidal crystals. In 2015, Zhang et al. [26] reviewed optical sensors based on photonic crystal cavity enhancing mechanisms. Overall speaking, the reported technologies can be considered in two categories. The first approach can be concluded as the "refractive index sensing", which could sensitively measure refractive index changes with gas involvement. Another one is known as "photonic crystal light absorption spectroscopy". This method is to detect the distinctive absorption patterns of gas molecules in the infrared spectrum. Because photonic crystals can slow light propagation and enhance light intensity in the space where gas fills [4], this new spectroscopy not only shares the advantages of the conventional spectroscopy but also eliminates the issues caused by large optical interaction length.

To the best of our knowledge, the research emphasis on the PhC-based gas sensing development has been mainly focused on the near-infrared spectral range. However, compared with the near-infrared region, molecular species in the mid-infrared range show intrinsic absorption bands with much larger absorption coefficients. **Figure 1** presents that the mid-infrared portion of the spectrum with several trace gas chemical species placed where their strong absorptions occur. As can be seen, taking the carbon dioxide (CO_2) as an example, its absorption strength in the mid-infrared range (~4.2 μm) is about two orders of magnetite higher than the one in near-infrared range (~2 μm). Such significant difference also exists in almost all the gas molecules including xylene, methane, and so on [29]. Therefore, fundamentally speaking, the optical sensors functioning in the mid-infrared range offer much higher device sensitivity [30]. Consequently, much richer information can be found for those wishing to probe, detect, image, or quantify these and many other species including explosives, nerve agents, and toxins [30]. Nevertheless, there are some hurdles preventing the development of PhC based optical sensors in

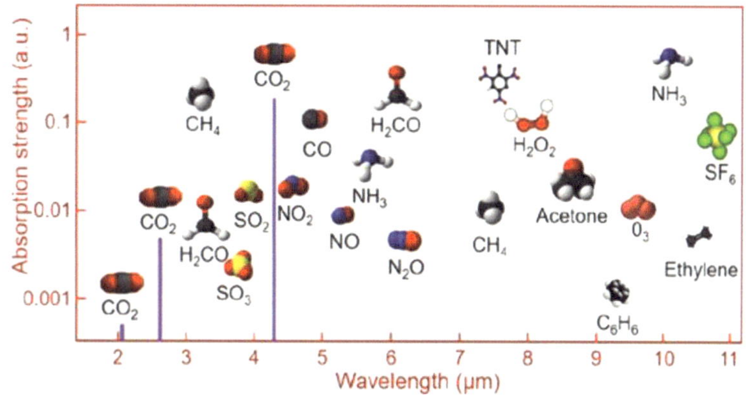

Figure 1.
The absorption strength of some typical trace gas molecules in the mid-infrared range [28].

the mid-infrared spectrum, such as the limited availability of the low-cost, high-efficient light sources or photodetectors. Also, it was pointed out in a report that the difficult process of alignment of the beam for coupling light in and out of the sample could also be very challenging, due to limitations of available equipment [31]. In order to have a systematic understanding of the current progress of PhC based gas sensing research in mid-infrared range, in this chapter, we are going to provide a comprehensive review on the existing mid-infrared PhC-based gas sensor technologies, evaluate their performance in a practical point of views, and also discuss the future of the PhC-based mid-infrared sensing technologies.

2. PhC-based mid-infrared gas sensing methods

It is known that most of the molecules have their distinctive absorption patterns in the mid-infrared spectrum, and as it was pointed out that their absorption coefficients are much higher than the ones in near-infrared range. Majorly due to these reasons, the most active research efforts of using PhC technology for gas sensing in mid-infrared range focuses on measuring the spectral intensity change caused by the gas resonant mode absorption mechanisms. But it is necessary to point out that, other than this typical methodology, some other approaches using PhC structures have also been reported, which rely on the detection of the peak position drifting of transmission or reflection spectrums caused by the gas-induced refractive index variation, or the electrical conductance behavior in the presence of gases. These methods certainly help to enrich the PhC sensing capabilities in the mid-infrared range. With all these being said, based on the different sensing mechanisms, in the following contents, we will elaborate on the recent development of the mid-infrared PhC based gas sensing technologies.

2.1 Mid-infrared light absorption sensing

Being as the most popular technology in the mid-infrared range, the gas absorption spectroscopy can be used to precisely measure both the gas composition and concentrations. In this method, the mid-infrared radiation is absorbed at some specific frequencies due to the presence of gas molecules. These frequencies correspond to vibrational modes of the molecular structures [32]. Typically, this type of gas sensor consists of three parts: the mid-infrared light source, the light/gas interaction chamber, and the radiation detector [33]. In recent years, PhC has been proven to be an effective technology to improve the performance of the light sources, especially for the laser devices [34]. But more significantly, PhC has also played a critical role to downsize the sensor module footprint by minimizing the interaction chamber volume under the chip scale level [35–38].

2.1.1 Advanced porous gas sampling structure for gas sensing

As proposed by Soref [29], silicon is a proper candidate for the mid-IR wavelength range due to its transparency window between 1.1 and 8.5 μm. While silicon dioxide is also transparent up to around 3.6 [29], silicon-on-insulator (SOI) is a suitable structural form for mid-IR integrated photonics [39]. In 2007, Lambrecht et al. [12] for the first time, suggested the implementation of a two dimensional (2D) macroporous silicon PhC in the interaction volume to slow the light and enhance the gas-light interaction. The experimental results with CO_2 showed more than two times enhancement in absorption line. In fact, these experimental results became a base for the realization of high sensitivity and miniature gas sensors. The device configuration is shown in **Figure 2**. Gas flows through the sampling cell from the top hole to the bottom, as highlighted in blue color. The sampling cell containing a PhC membrane is positioned between a thermal emitter and a pyrodetector with an IR bandpass filter centered at the absorption peak of CO_2 (λ = 4.24 μm). The PhC membrane is placed in between two BaF_2 light guiding rods which help to couple the IR radiation among the thermal emitter, PhC membrane and the pyrodetector. Here, one thing is worth noting that, the PhC membrane could be easily removed from the plastic holder without changing positions of the BaF_2 rods. This offers the possibility of measuring the empty cell with the same optical path length. In detail, the operation voltage is 10 Hz and the pyrodetector is measured by digital lock-in amplifier with a time constant of 2 s. With such setup, this method does offer the capability to detect the presence of CO_2 gas, which proves the possibility of using PhC technology for developing gas sensors with compact footprint. However, there is a crucial drawback with this simple PhC gas sensor. Typically, a low group velocity corresponds to a high effective refractive index, which leads to difficulties in and out couplings of radiation.

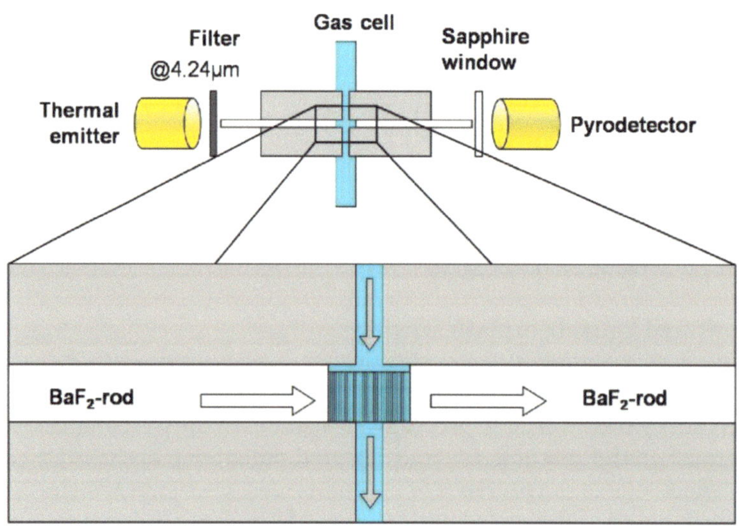

Figure 2.
Schematic design used for gas absorption measurements [35].

In 2011, Pergande et al. [35] solved this problem via designing antireflection layers (ARL) at two interfaces, including the air-ARL interface and the ARL-PhC interface. Because of that, it enhanced the light absorption of CO_2 up to 3.6 times. To explain the mechanism, introducing this ARL leads to the generation of surface modes which are used for coupling light into slow light PhC modes. These modes are confined in air-ARL interference due to the forbidden propagation of light in photonic band-gap. The thickness of ARL affects the spectral position of surface modes. When ARL thickness is equal to 0.57a (a is the lattice constant of PhC), the similarity of field distributions of surface modes and the slow light PhC modes would be able to couple together much easily. In addition, this ARL could also help to effectively reduce the interface optical loss. Theoretically, absorption of the device can be enhanced up to 60 by using this ARL enhanced PhC configuration. The difference between theoretical and experimental results comes from several factors, such as the non-optimal lattice constant and pore diameter fluctuation. They all could lead to the off-resonance operation mode to the CO_2 absorption line. In this work, 2D PhCs were fabricated by photoelectrochemical etching of n-type silicon. ARL is designed by photolithography and was transferred by photoelectrochemical etching into the silicon wafer. The lattice constant was 2 μm an (r/a), varying from 0.360 to 0.385. lengths of PhC changed from 100 μm to 1 mm while the height and width were approximately 330 μm and 1 mm, respectively. The porosity of the sample was about 64%. Pergande et al., according to numerical estimates, suggested that the positional variations and pore diameter fluctuations should be below 0.5% in order to allow for a reasonable transmission in the 1 mm device.

In the course of the experiment, several sample lengths, varying from 0.25 to 1 mm, were investigated. Results showed that their absorption enhancements changed at the range of 2.6–3.5. By taking all the mismatch factors into consideration, the numbers are in good agreement with the numerical simulation results. All the data are presented in the following **Table 1**.

In 2000, Boarino et al. [40] applied this porous silicon method to the environmental analysis of NO_2 gas because of its high sensitivity to surface molecular structures. **Figure 3** shows the enhanced absorption value due to the presence of NO_2. In this figure, spectrum-1 is related to the sample outgassed under dynamic vacuum, spectrum-2 is given after the dosage of 1 Torr of the pure NO_2 sample, and spectrum-3 is recorded after outgassing. As can be seen, in the range of 1000–1250 cm^{-1} (assigned to the stretching vibrations of Si–O species) and range of 2150–2300 cm^{-1} (Si–H stretching modes of SiO–H_x species) the absorption enhancement in the presence of NO_2 is intense while this value, at 3748 cm^{-1} (SiO–H stretching vibration), is negligible. They claimed that this broad absorption can be attributed to electrons populating of the conduction band. Moreover, from spectrum-3 is an evidence that removing NO_2 from the gas phase leads to a complete restoration of initial conditions. Moreover, the most recent research based on this method showed the enhanced gas absorption by a factor of 5.8 at 5400 nm [41].

It can be concluded that porous silicon is an effective gas sampling structure that can help in minimizing the overall size of the sensor device. However, the main

Length (mm)	Gas	Filter[μm/(a/λ)]	ζ_{exp}	$\zeta_{theo,TM}$	$\zeta_{theo,TE}$
1	CO_2	4.24/0.472	3.5	3.7	2.9
0.5	CO_2	4.24/0.472	2.6	3.7	2.9
0.5	CO_2	4.24/0.472	3.0	3.7	2.9

Table 1.
Experimental and theoretical value of the gas absorption enhancement [35].

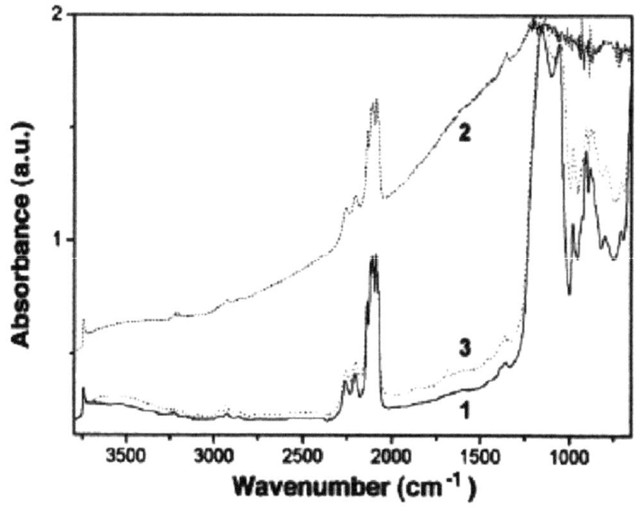

Figure 3.
FTIR spectra of free-standing p + PSL: Spectrum-1 is related to sample outgassed under dynamic vacuum. Spectrum-2 has been obtained after dosage of 1 Torr of pure NO_2, spectrum-3 has been recorded after outgassing the sample, under dynamic vacuum [40].

restrictive factor for using porous silicon PhCs in gas sensing is its high sensitivity to the small fluctuation of the pore diameter and the lattice constant. For instance, more than 0.5% pore diameter fluctuation in the 1 mm device eliminates the advantage of using the porous silicon. Thus, a high technology is needed for fabricating this porous silicon, which limits the implementation of this method. However, for cases where high-sensitive and small sensors are needed, using porous silicon PhCs to decrease the interaction path and increase the sensitivity is a suitable option.

2.1.2 Photonic crystal waveguide (PCW) for gas sensing enhancement

Mid-IR PhC waveguides (PCW) with high-quality factor are powerful tools for nonlinear optical applications because they can achieve slow-light enhancement and low linear propagation loss simultaneously [31]. Therefore, this component is of major interests in the gas sensing research area. The enhanced detection sensitivity achieved by using PCWs (strip and slot waveguides) in the near-infrared range have been previously reported [42, 43]. However, based on our knowledge, very few works about the PCW enhanced gas sensor in the mid-infrared range has been reported so far [44]. In 2015, Zou et al. [45] provided the first experimental demonstration of transmission characteristics of holey and slotted PCWs in silicon-on-sapphire at the wavelength of 3.43 μm with a fixed-wavelength inter-band cascade laser (ICL). They used an 800 μm long holey PCW to detect gas-phase Triethyl phosphate (TEP) with the concentration of 10 ppm (parts per million).

They investigated a holey PCW with a row of smaller holes (r_s = 0.65r), which was located in the center of PCW (hexagonal lattice of air holes in silicon) where air hole radius was r = 0.25*a* (*a* is lattice constant). As shown in **Figure 4** the optimization of (r_s/r) value has been conducted through considering four conditions: having a large guiding bandwidth for the propagating PCW modes, broad mode bandwidth, large electrical field overlap with the analyte, and high peak enhancement factor which yield to more efficient light-matter interaction. The other value that should be optimized is the lattice constant. As shown in **Figure 5** for a less than 830 nm, due to stopgap, there is no transmission. Moreover, when *a* is located between 840 and 845 nm the propagation loss is approximately 15 dB/cm, while

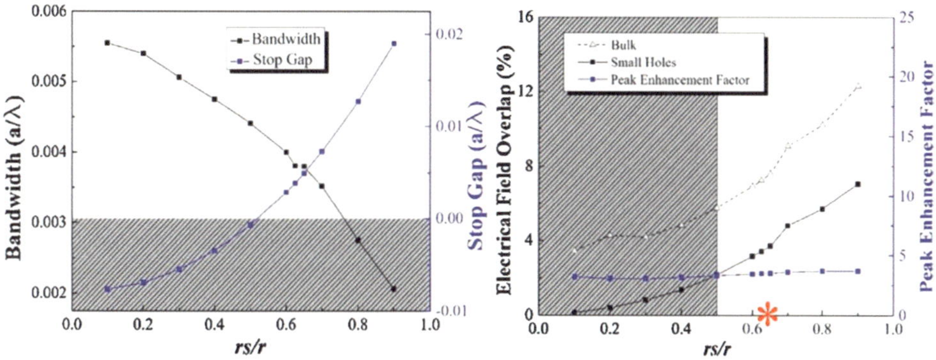

Figure 4.
Optimization of r_s/r according to bandwidth, stop gap electrical field overlap, peak enhancement factor [45].

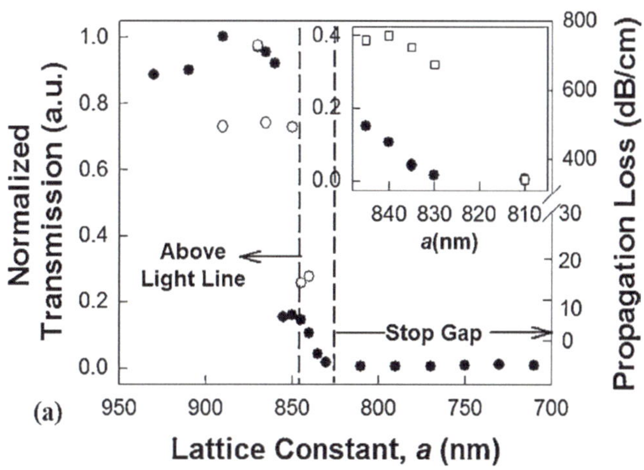

Figure 5.
Lattice constant dependent transmission and propagating loss [45].

this value increases dramatically for a > 845 nm. Therefore, it can be concluded that in order to have reasonable transmission and propagation loss, the lattice constant should have a value between dashed lines in **Figure 5**.

On the other hand, the slotted type PCW was also investigated. Likewise, the same optimization process was conducted for slotted PCW to determine slot width (130 nm) and lattice constant (between 830 and 840 nm). However, the optimized value showed propagation loss of 55 dB/cm, which is almost 3 times greater than the loss propagation in holey PCW. Afterward, they compared the peak of electric field enhancement factor in the holey PCW, and the conventional rectangular slotted PCW, to a regular PCW. This comparison showed 3.5 times and 13 times enhancement for holey and slotted PCW (respectively) relative to regular PCW. Furthermore, the electrical field overlap with the analyte in regular PCW was 5%, while this value for holey and slotted one was 8 and 15% respectively. However, propagation loss for holey PCW was 15 dB/cm while this value was 55 dB/cm for slotted one. Thus, we can conclude that the holey PCW can be a better candidate for absorption spectroscopic gas sensing because while its electrical field overlap with the analyte is 2 times lower than slotted PCW, the propagation loss of slotted PCW is 3 times higher than holey PCW.

The transmitted light through an 800 μm long holey PCW with a = 845 nm is measured in the presence and absence of TEP. They investigated changes in transmitted light intensity at λ = 3.43 μm for switching and steady state TEP flow through the holey PCW. For the switching flow, transmitted signal intensity dropped to 80% of its original intensity due to the presence of TEP. However, for the steady-state condition, this value dropped to 60% of its original value. Also, the measurements are independent of the flow rate of nitrogen at 10 and 50 ppm respectively. The noise in measurement comes from the electrical noise of detector and the vibration of the optical fiber. Furthermore, by replacing the slot and strip waveguide for the steady-state gas flow, instead of the holey PCW, just a small change in intensity of transmitted light at 3.43 μm is observed at 28 ppm TEP concentration. Thus, as **Figure 6** shows, it can be concluded that the holey PCW shows more sensitivity relative to the slot and strip waveguide because the holey PCW enhances both f (fill factor denoting relative fraction of optical field residing in the analyte medium) and group index, which is inversely related to group velocity. However, the slot waveguide only enhances the f factor, and has no impact on group index. Moreover, the high detection capability of PCWs in gas sensing is revealed in [46] where the detection of carbon monoxide with the concentration of parts-per-billion is possible.

In comparison to the porous silicon PhC Sensing method which enhances the sensor performance by only reducing group velocity, the PCW, especially the holey PCW, shows higher sensitivity because it enhances both f (filling factor denoting relative fraction of optical field residing in the analyte medium) and group velocity reduction simultaneously. In fact, using a PCW, instead of a simple 2D PhC, enables us to confine light and gas in at the same place, and increase the possibility of the interaction. Moreover, the propagation loss in PCWs is lower than the porous silicon PhC. However, the fabrication and optimization of a PCW are more complicated than the porous silicon PhC which potentially can lead to higher costs for PCW gas sensor development.

2.1.3 Mid-infrared PhC fiber enhanced sensing

Optical fibers offer significant advantages for gas sensing majorly due to its ability to confine optical radiation across long distances. It eliminates the need for beam collimation thus reduces device complexity significantly. However, their performance can be limited due to low mechanical flexibility, a weak overlap between light and gas or the requirement for conventional extrinsic gas cells [47]. Also, most of the fibers operate below 2 μm demonstrated due to the limited silica transmission window. In contrary to other microstructured fiber, Photonic bandgap fibers (PBFs) guide light in a gas (air) core rather than solid [48, 49]. This leads to the integration of gas sampling cell into the fiber, and the confinement of over 99% of the light in the hollow area rather than the silica, which makes PBFs an ideal

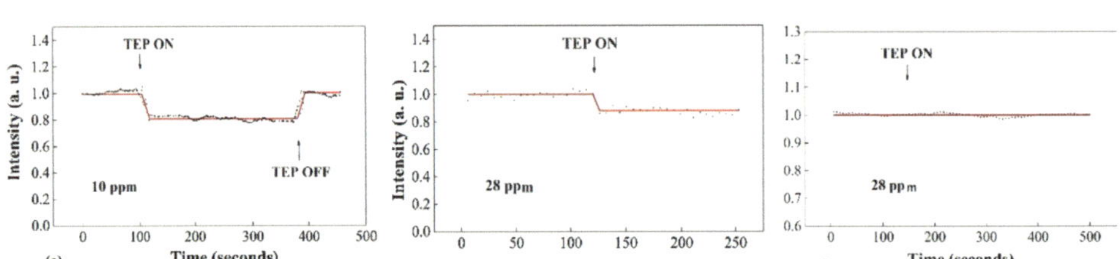

Figure 6.
(a). Change in transmitted light intensity through an 800 μm long holey PCW with a = 845 nm with 10 ppm TEP (b). Transmitted light intensity through a silicon slot waveguide in SoS in the presence and absence of 28 ppm TEP. (c). Transmitted light intensity through a silicon strip waveguide in SoS in the presence and absence of 28 ppm TEP [45].

candidate for miniaturized gas sensing applications [50, 51]. Furthermore, compared with the fused silica, these fibers can transmit light in the mid-infrared with a much lower optical loss (<1 dBm^{-1}) [52, 53].

Shephard et al. reported a bandgap guidance at 3.14 μm in a silica-based air-core photonic crystal fiber in 2005 [54]. The year after, they then investigate the gas sensing functionality using this PBF technology [53, 55]. The optical guiding range of this PBF was specially developed for methane sensing at ~3.2 μm [55]. The hollow-core PBF sensing unit was able to measure the methane vapor with the concentration of 1000 ppm. In the experiment, several fibers with a core diameter of ~45 μm and a pitch (distance between two neighboring cladding holes) between 7 and 8 μm were examined thoroughly. First, an 80 cm PBF was filled with the mixture of methane and nitrogen with the ratio of 5 (methane) to 95 (nitrogen) at the pressure of 2 bar. Then, the fiber was filled with nitrogen to normalize the measurements. The results are shown in **Figure 7** for 5, 1, 0.5, and 0.1% methane concentration. (solid curve) shows experimental results and the calculated absorption demonstrated by the dashed curve. Even in this low concentration (**Figure 7d**), the main absorption line of methane close to 3.32 μm is still visible, and the transmission dropped to significantly at this wavelength. Moreover, authors claimed that the small difference between the theoretical and experimental results is due to although processing errors during the spectral concatenation procedure.

In comparison to the porous silicon PhC as well as the PCW sensing mechanisms, the PBF gas sensor shows a higher energy overlap with gases and a lower optical loss. However, fibers are long in nature and this can be an obstacle to realization of SWaP sensors. Moreover, the degree of complexity for fabrication of PBFs can be lower than PCWs but is still higher compared to the porous silicon PhC devices.

2.2 Mid-infrared "refractive index" sensing

Besides of exploring the use of PhCs in the mid-infrared spectroscopy, there is another popular approach measuring the shift of PhC modulated Bragg resonant peak due to the refractive index change, which can lead to the detection of gas

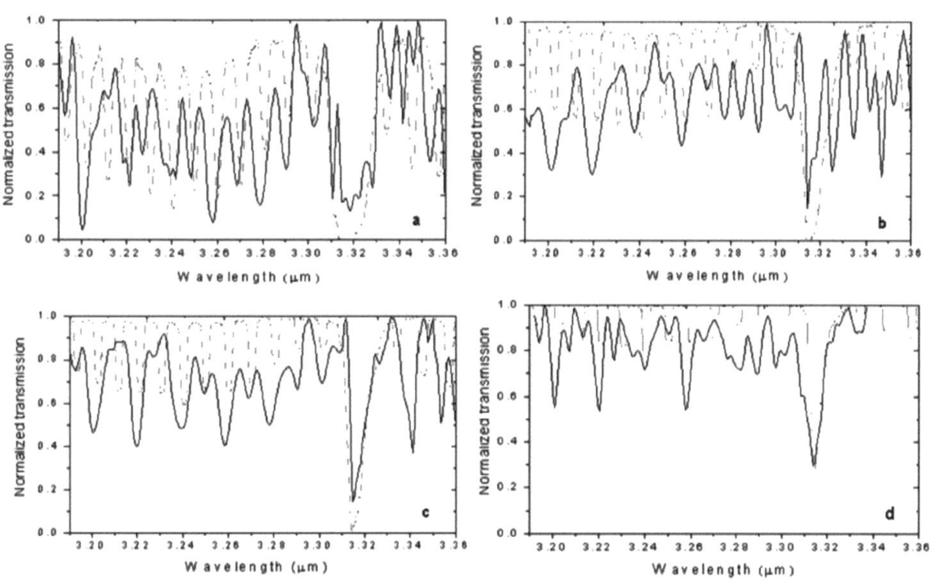

Figure 7.
Measured (solid curves) and theoretical (dash curves) absorption lines of methane for concentrations of a, 5%; b, 1%; c, 0.5%; and d, 0.1% [55].

concentrations. For instance, in a porous silicon PhC, when pore areas are filled by a gas the effective refractive index of the PhC will be increased. Moreover, it is possible that the lattice constant of the PhC increase due to the swelling of the gas or vapor. Understanding this procedure can be easier by considering Bragg law expression [27].

$$m\lambda = 2nd\,sin\theta \tag{1}$$

where m is the diffraction order, λ is the incident wavelength, n is the effective refractive index, d is the lattice constant, and θ is the glancing angle between the incident light and diffraction crystal planes [56, 57]. This method has been used in near and visible range [58–60], more frequently rather than mid-infrared range.

In 2015 Zou et al. [45] designed a holey, slotted, and a regular PCW for detection the chemical warfare simulant in the mid-infrared range. A section of this research has been assigned to the investigation of the relationship between the value of electrical field overlap with the analyte and the sensitivity of the refractive change based sensors. In order to study this correlation, they used 3D FDTD simulation. They selected C_2Cl_4 (refractive index = 1.5) as a top cladding. As shown in **Figure 8** the sensitivity of a PCW sensor is depended directly on the percent of electrical field overlap with the analyte. So that, the shifted transmission in slotted PCW is 2 times higher than holey one because its electrical field overlap with the analyte is almost 2 times higher as well. In 2017 Turduev et al. [61] presented an optical refractive index sensor (T-slotted PC sensor) design for mid-IR photonics. They used numerical methods based on finite-difference time-domain and plane-wave expansion method. an overall sensitivity is calculated to be around 500 nm/RIU for the case of higher refractive indices of analytes n = 1.10–1.30.

Detecting an unknown gas through this method can be challenging because the principles of this method are based on the refractive index changes. Thus, detecting two different gases with the same refractive index is simply impossible. On the other hand, for gases with refractive indexes close to air refractive indexes (n = 1), the sensitivity of this kind of sensors can be strongly reduced. However, the complexity of this method is lower than absorption-based sensing methods.

2.3 Mid-infrared electrical conductance sensing

In 2000 Boarino et al. [38] studied changes in electrical conductivity in the presence of NO_2 through P^+ porous silicon layers (PSL) at the room temperature and the atmospheric pressure. The recovery time and response to interfering gases were tested as well. PSL is well-known in humidity sensors area, while its application in the field of gas sensing has been considered only recently. This structure has obtained high importance in the field of gas sensors due to its high surface to

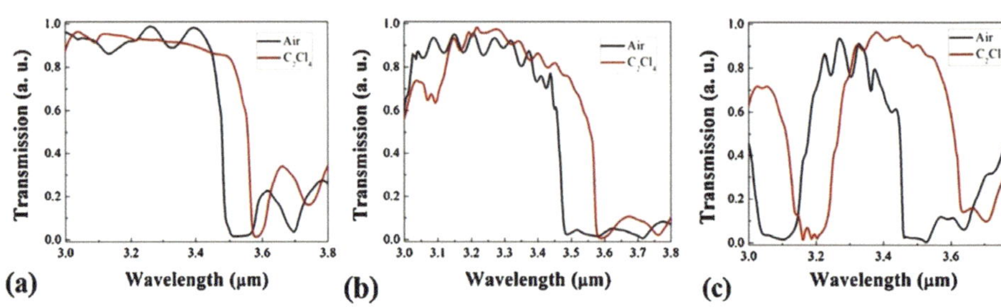

Figure 8.
Simulated transmission with air-clad and C_2C_{l4}-clad conditions for (a) conventional PCW, (b) holey PCW, and (c) slotted PCW [45].

Porosity (%)	ΔG/G (3 ppm)	ΔG/G (5 ppm)	ΔG/G (10 ppm)
38	6	87,837	0.7
43	1	1.9	3.5
53	0.3	1.7	4.6
62	29.3	84.3	197
75	3.5	45	164

Table 2.
Relative response of PS samples of different porosity to the listed NO_2 concentrations [38].

volume ratio, and its reactivity to the environment. Changes in work function, refractive index, photoluminescence, and conductivity variation can be used as indices in the gas sensing. Boarino et al. used the last feature to detect NO_2. Accordingly, due to the inherent characteristic of P^+ mesoporous silicon, strong changes can be observed in resistivity in the presence of polar liquids, vapors, and gases. They measured the PS change of conductance in presence of different gases in humid conditions at constant bias V = 5 v. The PS response to NO_2 was tested for different value of porosity. **Table 2** shows the relative conductance variation (ΔG/G) for different porosity (38, 43, 53, 62 and 75%).

Since the high available surface for gas adsorption plays a key role in obtaining an efficient chemical sensor, the 55% porosity for PS surface gives the best result, due to the maximum value of surface to volume ratio. The lowest concentration that could be examined, was 1 ppm and the relative response was 1.6 for the 60% porosity sample. In comparison with NO_2, under the same concentration, the conductivity change in the presence of NO was significantly lower. Likewise, the relative response of PS to interfering species (CO (up to 1000 ppm) and CH_4 (up to 5000 ppm)) and alcohol, such as methanol, at concentrations up to 800 ppm was negligible. They also used FTIR spectroscopy for detecting NO_2 through PSL which is studied in 2.1.1 section in details.

3. Conclusion

In this chapter, we presented a review work on the recent progress of PhC-based gas sensing research in the mid-infrared range. Various material structures including using porous silicon structure, photonic crystal waveguides, and hollow-core photonic crystal fibers, as well as both optical and electrical detection methods, have been thoroughly discussed. As mentioned, porous silicon structure enhanced sensing device achieved the highest sensitivity to detect NO_2 at 1 ppm concentration level through measuring the conductance changes. But this method is restricted to a limited range of gases, and is unable to detect nonpolar gases such as CO, CH_4, and alcohols. The other issue can be related to electrical components which are necessary for this method. These electrical components increase the risk of electrical discharge and augment the risk of explosion. Moreover, the electrical noises can strongly affect this kind of sensors. For the holey PCW, the sensor unit can deliver the measurement of Triethyl phosphate (TEP) with the concentration of 10 ppm. The small size (800 μm) of this PCW offers a great advantage which can potentially lead to the realization of SWaP sensors. The main drawback of this kind of sensors is that they are so sensitive to small fluctuation in the hole diameter. Thus, the fabrication process for this kind of sensors might be difficult and time-consuming. However, the high energy overlap with gases within the holey PCWs, and its high power in slowing light and its small size make this sensor one of the best candidates for gas sensing applications. While, compared to the PCW, the PBF shows more electrical overlap with gases and lower propagation loss. Generally speaking, the development of PhC-based mid-infrared gas sensing research is still in its early stage and not ready for commercialization. However, considering the strong demands from IoT infrastructure for modern sensor units with combined SWaP-C features, it is anticipated that the progress in the PhC-based mid-infrared gas sensing area will develop much faster in the coming years, and some of the discussed technical approaches could eventually advance to a practical level and make a significant impact to our daily life.

Author details

Tahere Hemati and Binbin Weng*
School of Electrical and Computer Engineering, University of Oklahoma, Norman, OK, United States

*Address all correspondence to: binbinweng@ou.edu

References

[1] Azad A, Akbar S, Mhaisalkar S, Birkefeld L, Goto K. Solid-state gas sensors: A review. Journal of the Electrochemical Society. 1992;**139**(12):3690-3704

[2] Eranna G, Joshi B, Runthala D, Gupta R. Oxide materials for development of integrated gas sensors a comprehensive review. Critical Reviews in Solid State and Materials Sciences. 2004;**29**(3-4):111-188

[3] Shimizu Y, Egashira M. Basic aspects and challenges of semiconductor gas sensors. MRS Bulletin. 1999;**24**(6):18-24

[4] Jaaniso R, Tan OK. Semiconductor Gas Sensors. Amsterdam: Elsevier; 2013

[5] Krebs P, Grisel A. A low power integrated catalytic gas sensor. Sensors and Actuators B: Chemical. 1993;**13**(1-3):155-158

[6] Spetz A, Winquist F, Sundgren H, Lundstrom I. Field effect gas sensors. In: Gas Sensors. University of Brescia, Italy: Springer; 1992. pp. 219-279

[7] Schmidt JC, Campbell DN, Clay SB. Electrochemical Gas Sensor. June 25 1985. US Patent 4,525,266

[8] Hodgkinson J, Tatam RP. Optical gas sensing: A review. Measurement Science and Technology. 2012;**24**(1):012004

[9] Sberveglieri G. Gas Sensors: Principles, Operation and Developments. University of Brescia, Italy: Springer Science & Business Media; 2012

[10] Gubbi J, Buyya R, Marusic S, Palaniswami M. Internet of things (IoT): A vision, architectural elements, and future directions. Future Generation Computer Systems. 2013;**29**(7):1645-1660

[11] Jones KH, Gross JN. Reducing size, weight, and power (swap) of perception systems in small autonomous aerial systems. In: 14th AIAA Aviation Technology, Integration, and Operations Conference. 2014. p. 2705

[12] Lambrecht A, Hartwig S, Schweizer S, Wehrspohn R. Miniature infrared gas sensors using photonic crystals. In: Photonic Crystal Materials and Devices VI. Vol. 6480. San Jose, California, United States: International Society for Optics and Photonics; 2007. p. 64800D

[13] Gao Y, Shiue RJ, Gan X, Li L, Peng C, Meric I, et al. High-speed electro-optic modulator integrated with graphene-boron nitride heterostructure and photonic crystal nanocavity. Nano Letters. 2015;**15**(3):2001-2005

[14] Ge X, Shi Y, He S. Ultra-compact channel drop filter based on photonic crystal nanobeam cavities utilizing a resonant tunneling effect. Optics Letters. 2014;**39**(24):6973-6976

[15] Fasihi K. High-contrast all-optical controllable switching and routing in nonlinear photonic crystals. Journal of Lightwave Technology. 2014;**32**(18):3126-3131

[16] Lin CY, Subbaraman H, Hosseini A, Wang AX, Zhu L, Chen RT. Silicon nanomembrane based photonic crystal waveguide array for wavelength-tunable true-time-delay lines. Applied Physics Letters. 2012;**101**(5):051101

[17] Yablonovitch E. Inhibited spontaneous emission in solid-state physics and electronics. Physical Review Letters. 1987;**58**(20):2059

[18] John S. Strong localization of photons in certain disordered dielectric superlattices. Physical Review Letters. 1987;**58**(23):2486

[19] Joannopoulos JD, Johnson SG, Winn JN, Meade RD. Photonic Crystals: Molding the Flow of Light. Princeton, New Jersey: Princeton University Press; 2011

[20] Notomi M. Strong light confinement with periodicity. Proceedings of the IEEE. 2011;**99**(10):1768-1779

[21] Yablonovitch E. Inhibited spontaneous emission in solid-state physics and electronics. Physics Review Letters. 1987;**58**:2059

[22] John S. Strong localization of photons in certain disordered dielectric superlattices. Physics Review Letters. 1987;**58**:2486

[23] Li H, Chang L, Wang J, Yang L, Song Y. A colorful oil-sensitive carbon inverse opal. Journal of Materials Chemistry. 2008;**18**(42):5098-5103

[24] Snow P, Squire E, Russell PSJ, Canham L. Vapor sensing using the optical properties of porous silicon Bragg mirrors. Journal of Applied Physics. 1999;**86**(4):1781-1784

[25] Lee K, Asher SA. Photonic crystal chemical sensors: Ph and ionic strength. Journal of the American Chemical Society. 2000;**122**(39):9534-9537

[26] Zhang YN, Zhao Y, Lv RQ. A review for optical sensors based on photonic crystal cavities. Sensors and Actuators A: Physical. 2015;**233**:374-389

[27] Xu H, Wu P, Zhu C, Elbaz A, Gu ZZ. Photonic crystal for gas sensing. Journal of Materials Chemistry C. 2013;**1**(38):6087-6098

[28] Nicoletti S. Mid-Infrared Photonics Devices Fabrication for Chemical Sensing and Spectroscopic Applications. Alpexpo in Grenoble, France: Semicon Europa; 2016

[29] Soref R. Mid-infrared photonics in silicon and germanium. Nature Photonics. 2010;**4**(8):495

[30] Seddon AB. A prospective for new mid-infrared medical endoscopy using chalcogenide glasses. International Journal of Applied Glass Science. 2011;**2**(3):177-191

[31] Reimer C, Nedeljkovic M, Stothard DJ, Esnault MO, Reardon C, O'Faolain L, et al. Mid-infrared photonic crystal waveguides in silicon. Optics Express. 2012;**20**(28):29361-29368

[32] Stuart B. Infrared Spectroscopy. Sydney, Australia: Wiley Online Library; 2005

[33] Korotcenkov G. Handbook of Gas Sensor Materials: Properties, Advantages and Shortcomings for Applications. Vol. 2. Gwangju, South Korea: New Trends and Technologies; 2014

[34] Hvozdara L, Gianordoli S, Strasser G, Schrenk W, Unterrainer K, Gornik E, et al. Spectroscopy in the gas phase with gaas/algaas quantum-cascade lasers. Applied Optics. 2000;**39**(36):6926-6930

[35] Pergande D, Geppert TM, Rhein Av, Schweizer SL, Wehrspohn RB, Moretton S, et al. Miniature infrared gas sensors using photonic crystals. Journal of Applied Physics. 2011;**109**(8):083117

[36] Jensen KH, Alam M, Scherer B, Lambrecht A, Mortensen NA. Slow-light enhanced light-matter interactions with applications to gas sensing. Optics Communications. 2008;**281**(21):5335-5339

[37] Conteduca D, Dell'Olio F, Ciminelli C, Armenise M. New miniaturized exhaled nitric oxide sensor based on a high q/v mid-infrared

1dphotonic crystal cavity. Applied Optics. 2015;**54**(9):2208-2217

[38] Chen Y, Lin H, Hu J, Li M. Heterogeneously integrated silicon pho-tonics for the mid-infrared and spectroscopic sensing. ACS Nano. 2014;**8**(7):6955-6961

[39] Milosevic MM, Nedeljkovic M, Ben Masaud TM, Jaberansary E, Chong HM, Emerson NG, et al. Silicon waveguides and devices for the mid-infrared. Applied Physics Letters. 2012;**101**(12):121105

[40] Boarino L, Baratto C, Geobaldo F, Amato G, Comini E, Rossi A, et al. No_2 monitoring at room temperature by a porous silicon gas sensor. Materials Science and Engineering B. 2000;**69**:210-214

[41] Kraeh C, Martinez-Hurtado J, Popescu A, Hedler H, Finley JJ. Slow light enhanced gas sensing in photonic crystals. Optical Materials. 2018;**76**:106-110

[42] Zou Y, Chakravarty S, Xu X, Lai W, Chen RT. Silicon chip based near-infrared and mid-infrared optical spectroscopy for volatile organic compound sensing. In: Lasers and Electro-Optics (CLEO), 2014 Conference on. San Jose, CA, USA: IEEE; 2014. pp. 1-2

[43] Lai WC, Zou Y, Chakravarty S, Zhu L, Chen RT. Comparative sensitivity analysis of integrated optical waveguides for near-infrared 14 volatile organic compounds with 1ppb detection. In: Silicon Photonics IX. Vol. 8990. San Francisco, California, United States: International Society for Optics and Photonics; 2014. p. 89900Z

[44] Ranacher C, Consani C, Tortschanoff A, Jannesari R, Bergmeis-ter M, Grille T, et al. Mid-infrared absorption gas sensing using a silicon strip waveguide. Sensors and Actuators A: Physical. 2018;**277**:117-123

[45] Zou Y, Chakravarty S, Wray P, Chen RT. Mid-infrared holey and slotted photonic crystal waveguides in silicon-on-sapphire for chemical warfare simulant detection. Sensors and Actuators B: Chemical. 2015;**221**:1094-1103

[46] Rostamian A, Guo J, Chakravarty S, Chung C-J, Nguyen D, Chen RT. Parts-per-billion carbon monoxide sensing in silicon-on-sapphire mid-infrared photonic crystal waveguides. In: CLEO: Applications and Technology. San Francisco, California, United States: Optical Society of America; 2018. p. ATh1O–4

[47] Rajan G. Optical Fiber Sensors: Advanced Techniques and Applications. Vol. 36. Vancouver, British Columbia, Canada: CRC Press; 2015

[48] Knight JC, Broeng J, Birks TA, Russell PSJ. Photonic band gap guidance in optical fibers. Science. 1998;**282**:1476-1478

[49] Smith CM, Venkataraman N, Gallagher MT, Müller D, West JA, Borrelli NF, et al. Low-loss hollow-core silica/air photonic bandgap fibre. Nature. 2003;**424**:657-659

[50] Roberts PJ, Couny F, Sabert H, Mangan BJ, Williams DP, Farr L, et al. Ultimate low loss of hollow-core photonic crystal fibres. Optics Express. 2005;**13**:236-244

[51] Saini TS, Kumar A, Sinha RK. Broadband mid-infrared super-continuum spectra spanning 2-15 μm using as 2 se 3 chalcogenide glass triangular-core graded-index photonic crystal fiber. Journal of Light wave Technology. 2015;**33**(18):3914-3920

[52] Russell PSJ. Photonic-crystal fibers. Journal of Lightwave Technology. 2006;**24**:4729-4749

[53] Gayraud N, Stone JM, MacPherson WN, Shephard JD, Maier RRJ, Knight JC, et al. Mid infrared gas sensing using a hollowcore photonic bandgap fibre. In: Optical Fiber Sensors, OSA Technical Digest (CD). Cancun Mexico: Optical Society of America; 2006. paper ThA5

[54] Shephard JD, MacPherson WN, Maier RR, Jones JDC, Hand DP, Mohebbi M, et al. Single-mode mid-IR guidance in a hollowcore photonic crystal fiber. Optics Express. 2005;**13**:7139-7144

[55] Gayraud N, Kornaszewski LW, Stone JM, Knight JC, Reid DT, Hand DP, et al. Mid-infrared gas sensing using a photonic bandgap fiber. Applied Optics. 2008;**47**(9):1269-1277

[56] Bragg WH, Bragg BWL, et al. The reflection of x-rays by crystals. Proceedings of the Royal Society of London A. 1913;**88**(605):428-438

[57] Asher SA, Holtz J, Liu L, Wu Z. Self-assembly motif for creating submicron periodic materials. Polymerized crystalline colloidal arrays. Journal of the American Chemical Society. 1994;**116**(11):4997-4998

[58] Kobler J, Lotsch BV, Ozin GA, Bein T. Vapor-sensitive bragg mirrors and optical isotherms from mesoporous nanoparticle suspensions. ACS Nano. 2009;**3**(7):1669-1676

[59] Yang H, Jiang P. Macroporous photonic crystal-based vapor detectors created by doctor blade coating. Applied Physics Letters. 2011;**98**(1):011104

[60] Lu G, Farha OK, Kreno LE, Schoenecker PM, Walton KS, Van Duyne RP, et al. Fabrication of metal-organic framework-containing silica-colloidal crystals for vapor sensing. Advanced Materials. 2011;**23**(38):4449-4452

[61] Turduev M, Giden IH, Babayigit C, Hayran Z, Bor E, Boztug C, et al. Mid-infrared T-shaped photonic crystal waveg-uide for optical refractive index sensing. Sensors and Actuators B: Chemical. 2017;**245**:765-773

Chapter 6

Modelling of Photonic Crystal (PhC) Cavities: Theory and Applications

Ahmad Rifqi Md Zain and Richard M. De La Rue

Abstract

In recent years, many researchers have shown their interest in producing a compact high-performance optical chip that is useful for most telecommunication applications. One of the solutions is by realising photonic crystal (PhC) structures that exhibit high-quality factors in a small mode volume, V. Silicon on insulator (SOI) is one of the main contenders due to its high-index contrast between the silicon (Si) core waveguide with silica (SiO_2) cladding surrounding it. The maturity of silicon photonic can also be incorporated with CMOS chips making it a desired material. A strong optical confinement provided by PhC structures makes it possible to realise the compact device on a single chip. In this chapter, we will discuss a fundamental background of photonic crystal cavities mainly on one-dimensional (1D) structures, which are the simplest as compared to their counterparts, 2D and 3D PhC device structures. We have modelled a photonic crystal cavity using finite-difference time-domain (FDTD) approach. This approach uses time-dependent Maxwell equation to cover wide frequency range in a single simulation. The results are then compared with the actual measured results showing a significant agreement between them. The design will be used as basic building block for designing a more complex PhC structures that exhibit high-quality factors for applications such as filtering, DWDM and sensors.

Keywords: photonic crystal (PhC), photonic wire (PhW), integrated optics, high-quality factors, finite-difference time-domain (FDTD)

1. Introduction

The potential importance of integrated optics was not fully realised until 1968. Light propagation in thin films has been proposed and developed extensively since then [1]. The term integrated optics relates to a wide variety of structures where the propagation of light is controlled by a thin dielectric film or by strips of dielectric. The range of laser frequencies available and the types of material used have their limitations. Initially, gas laser and solid-state lasers were used as the light sources in early experiments. There is a possible need for much smaller sources that can be used to achieve the requirement of integrated optics in order to integrate with other applications. For wavelengths smaller than a certain value, say around 0.1 μm, overcoming large absorption and scattering losses becomes a priority since the smaller wavelength range imposes limitations on the practical use of waveguiding.

In addition, waveguide integrated optics is based on electromagnetic waveguiding at optical frequencies using thin-film optics. In recent years, semiconductor devices have played a major role in the evolution of integrated optics, due to their significant properties relevant to the goal of monolithic integrated optical circuits. In the early 1960s, research on thin-film phenomena became the key route towards developing more complex waveguide properties. The guiding action of planar layers in p-n junctions was observed and reported in 1963 by Yariv and Leite [2] and Bond et al. [3]. Their result has been subsequently used by Nelson and Reinhardt [4] in providing light modulation via the electro-optic effect. Although there was no concern with the optical waveguide circuitry, this work was just the beginning of the new era of planar thin-film waveguides. Light propagation in thin films has been proposed and developed since then [5]. The subject of dielectric periodic microstructures has become a priority ever since the evolution of lasers and integrated optics generally in the early 1960s [5–11]. This great evolution was just the beginning of the new era of development of photonic microstructures on single compact chips. Much research has been carried out with the aim of providing faster optical communication and data processing—whether for entertainment, route switching or computational purposes. In recent years, the motivation towards producing compact and faster communication has become a platform for much research, including switching purposes.

In addition, the advance in photonic technology for many applications has emerged on a large scale, whether using active devices such as III–V semiconductor materials or even silicon and silica passive devices. But the latter two materials still work as a separate system, although the main aim is still to achieve a monolithic photonic integration that is capable of handling any application in a single chip. The developments based on the concepts formulated by Purcell [12, 13] regarding the effect of radiation properties due to the presence of mirrors have been discussed extensively. These ideas led to the new concept of photonic crystals (PhCs) [14, 15]. Instead of manipulating the electrons that are involved in the use of the conventional electronic properties of solids, where they can produce an electronic band gap, photons are manipulated in periodic structures (photonic crystals)—and can exhibit stopband and photonic band-gap behaviour. In other words, photons are not allowed to propagate through the 'crystal' structures at all—and there can be a forbidden gap or band gap.

Much of the attraction in the research areas of the micro- and nanophotonic structures comes from the use of high refractive index contrast materials such as silicon-on-insulator (SOI) that have been increasingly used in recent years. This development is due to the ability of silicon technology to support monolithic integration of optical interconnects and form fully functional photonic devices that can be incorporated into CMOS chips. Soref and Lorenzo [16] have demonstrated the possibilities of passive and active silicon waveguides as long ago as 1985, with single-crystal silicon grown epitaxially on a heavily doped silicon substrate. The advances of silicon-based and silicon-on-insulator optoelectronics have also been noted by Jalali et al. [17] and Masini et al. [18].

2. Photonic crystal and photonic wire waveguides

The concept of photonic band-gap structures was independently proposed by John [14] and by Yablonovitch [15]. PBG structures create the condition where over a certain photon energy range, light can travel through the periodic structure—and is reflected back when impinging onto the crystal and is not allowed to propagate—thus creating a so-called forbidden zone. In 1991, the first experimental demonstrations have shown that by using an array of holes drilled into the high refractive index material, a stopband is produced where no transmission is allowed over this frequency range [19]. Full PBG structures consist of three-dimensionally periodic structures that inhibit spontaneous emission within the electromagnetic band gap. New design has been developed and innovated based on this concept ever since, although improving the overall performance of this device is still a major concern for full device functionality—and there is also a performance limitation determined by various fabrication processes [19, 20]. 3D photonic crystal structures (PhCs) are one of the possible contenders for the provision of highly compact devices on a single chip that will allow the realisation of complex subsystems. Due to the inherent difficulties of realising and controllably modifying 3D structures, work on 2D and 1D structures has emerged tremendously—which is partly due to the lesser design complexity and the reduction in size. But they can produce some interesting results that have contributed significantly towards the realisation of photonic integrated circuits (PICs). The motivation towards miniaturising PIC devices has expanded the need to put more effort into designing compact photonic crystal-based devices. The massive development of telecommunication infrastructures has created a large demand of multiple applications on a single chip by using a combination of several optical subsystems.

In general, photonic crystal device structures exhibit a strong optical confinement covering a fairly large frequency spectrum. Strong optical confinement is needed in a small volume to provide a suitable platform in the optical emission properties—thus creating enhancement of the luminescent 'atoms' through spontaneous emission. By creating a 'defect' or a small region surrounded by the photonic crystal arrangement, the basic properties of that photonic crystal lattice are significantly changed. In other words, the photonic crystal has the capability to localise light when a 'defect' is introduced within the periodicity of the crystal arrangement —thus forming a micro-cavity that is surrounded by a highly reflective mirror region. For example, in 2D photonic crystal structures, a 'defect' or micro-cavity can be formed by simply removing one or more holes [21–23]—or by changing the surrounding hole sizes [24–26]. Light that is strongly confined within the channel waveguide formed by the photonic crystal arrangement (square [27] or hexagonal lattice [28, 29] is directly coupled into the micro-cavity region. In this design arrangement, light may be guided through the structure by removing a single row of holes to form a channel waveguide—and in this way, light can propagate at the characteristic frequency of the cavity, within the band gap. Channel waveguides may be designed to have different widths, W, such as W1 [30]—where a single row of holes is removed to provide a channel waveguide. In other examples, W3 [31] consists of three hole removed and W0.7 [32]—i.e. a situation where the spacing between two blocks of photonic crystal is additionally increased by 0.7 of a lattice spacing. Recently, low propagation losses, 4.1 dB/cm, have been obtained in a single-line defect W1 PhC channel waveguide [33] which shows that PhCs can provide a suitable platform for designing low loss devices. In 1D photonic crystal structures, the micro-cavity has great potential for producing a high-quality factor in a small volume—thus providing a suitable platform to design a wavelength selective device, for example, for WDM applications using passive components such as multiplexers/de-multiplexers, optical switching, sensors and optical filters. On the other hand, in 1D photonic crystals, micro-cavities may be formed by creating a defect and using a smaller hole in the middle of a single-row crystal, as shown in **Figure 1**.

In photonic crystal (PhC) micro-cavity structures, the optical properties may be characterised by the Q/V ratio (often called the Purcell factor [12, 13]), where Q is the quality factor and V is the modal volume corresponding to the particular

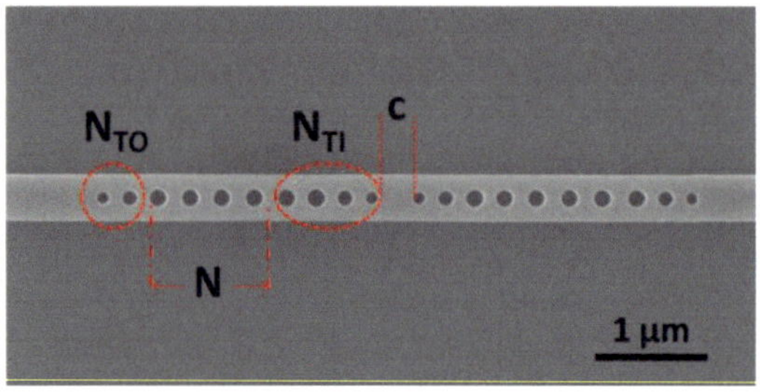

Figure 1.
1D PhC/PhW waveguide structures with a series of PhC hole of periodic spacing, a, and hole diameter, d, embedded in 500 nm wire. The tapered hole introduced has a number of hole tapered outside cavity, N_{TO}*, and the number of tapered hole within cavity,* N_{TI}*, with cavity length, c. A periodic mirror has N number of equally spaced hole.*

micro-cavity and its characteristic electromagnetic resonant modes. Thus designing high Q-factor optical micro-cavities confined in a small volume, V, may be useful for high-speed optical processing—where light is confined within a small volume on the order of $(\lambda/2n)^3$—and λ is the emission wavelength and n is the refractive index of the given material. Recently, designing ultrasmall micro-cavity devices based on 1D PhC/PhW dimensions has been of major interest because of their capability to provide extremely high Q/V values, close to the theoretical values of the modal volume, $V = (\lambda/2n)^3$. Q-factor values as large as 10^8 have been achieved, but this experimental value was based on silica toroids [34]—but this design has a relatively large modal volume, corresponding to a Q/V value of approximately 5×10^4 $(\lambda/2n)$.

Therefore, in most optical telecommunication applications, there is a need to have 1D PhC/PhW device structures that are necessary for manipulation of light at the infrared wavelength (around 1550 nm)—ruled by its capability of confined light within a small volume, V. Due to the fabrication challenges and the capability of designing structures that occupy very small areas, one-dimensional PhC structures have been preferred, although there are practical performance limitations. The devices typically consist of a single row of holes embedded in a narrow single-mode photonic wire waveguide. On the other hand, photonic wire (PhW) device structures based on total internal reflection (TIR) concepts have shown a capability for reduced loss, together with less complexity. They can also provide strong optical confinement due to the large refractive index contrast between the waveguide core and its surrounding cladding, leading also to small device volumes and compact structures [35]. In addition the photonic wire approach also gives great flexibility for the design of structures such as sharp bends, abrupt Y-junctions, small device volumes, micro-cavities and Mach-Zehnder (MZ) structures [36–42]. In other words, this concept is based on high refractive index contrast where light is confined in such a narrow ridge waveguide. The combination of one-dimensional photonic crystal (PhC) structures and photonic wire (PhW) waveguides in high refractive index materials such as silicon-on-insulator (SOI) became increasingly important in a number of research areas. In order to obtain a wide range of device functionality, the reduction of propagation losses in narrow photonic wires is equally as important as enhancing the performance of the device structures.

On the other hand, 1D PhC/PhW device structures have increasingly became a topic of interest—and use a mirror design that is based on a single periodic row of holes embedded in a narrow ridge waveguide, as shown in **Figure 1**. This approach was first introduced by Foresi et al. [43]. The periodic hole mirror characteristics can be varied by changing several parameters—such as hole diameter, cavity spacing and hole spacing—as will be described in detail in Chapter 4. In the present work, light confined within a PhC/PhW structure is directly coupled into the micro-cavity by using a tapered hole arrangement, as shown in **Figure 1**.

These kinds of device structure also have the capability of providing compact structures in small device volumes, as compared to other more complex structures such as ring resonators—which occupy larger device volumes. Furthermore, 1D PhC/PhW structures may also be preferred, since they can exhibit large band gaps as compared to what 3D photonic crystal structures can offer—thus making the PhC/PhW approach a contender for filter devices that can be integrated with other photonic devices. The fact that they share similar concepts with grating filters helps in understanding how these devices operate. With the material properties of SOI, an extremely small waveguide working in single-mode operation can be realised, with a reduction in propagation losses from 50 dB in 1996 [44] to 1.7 dB/cm in late 2006 [45], and most recently a propagation loss value of 0.91 dB/cm [46] has been achieved. The other features that have led to increasing attention to this area of research are the ability to provide a platform for the confinement of light within a small volume—for example, when a defect or a spacer is introduced between periodic mirrors. With this condition, light can be trapped within a small cavity, thus producing resonances that occur at certain frequencies within the stopband. These structures have been characterised by their high Q-factor value, adequate normalised optical transmission and small modal volume.

Maxwell's equations are important for an understanding of light propagation in photonic crystals. They are central for the solution of electromagnetic problems in dielectric media—for a variety of different lengths and dielectric scales, which are related to each other.

3. Photonic crystals: the theory

In photonic crystals, the famous Maxwell's equations are used to study light propagation in photonic crystal structure. The propagation of light in a medium is governed by the four well-known microscopic Maxwell's equations, written here in cgs units [21, 47, 48]. The microscopic forms of the Maxwell equation are given by

$$\nabla \cdot B = 0 \tag{1}$$

$$\nabla \cdot D = 4\pi\rho \tag{2}$$

$$\nabla \times E + \frac{1}{c}\left(\frac{\partial B}{\partial t}\right) = 0 \tag{3}$$

$$\nabla \times H - \frac{1}{c}\left(\frac{\partial D}{\partial t}\right) = \frac{4\pi}{c} J \tag{4}$$

or in mks/SI unit they can be written as

$$\nabla \cdot B = 0 \tag{5}$$

$$\nabla \cdot D = \rho \tag{6}$$

$$\nabla \times E + \left(\frac{\partial B}{\partial t}\right) = 0 \tag{7}$$

$$\nabla \times H - \left(\frac{\partial D}{\partial t}\right) = J \tag{8}$$

Based on Joannopoulos and Jackson [21, 47], Eqs. (1)–(4) are given in cgs units, whereas Eqs. (5)–(8) are given in mks/SI units, where the physical quantities are given as

B magnetic flux density in Tesla, T; *D* electric flux density in Coulombs per square m, C/m^2; *E* electric field strength in Volt per metre, V/m; *H* magnetic field strength in Ampere per metre, A/m; ρ electric charge density in Coulombs per cubic metre, C/m^3; J electric current density in Ampere per square metre, A/m^2.

The detailed derivation of each counterpart of Maxwell's equations is given by Jackson in Ref. [47]. For propagation in mixed dielectric medium, ρ and J are set to zero, since there are no free charges or currents in the homogeneous dielectric material. By assuming that the applied field strength is small and behave linearly, the dielectric flux density, D, can be related to the electric field density by the power series of

$$D_i = \sum_j \varepsilon_{ij} E_j + \sum_j k\chi_{ijk} E_j E_k + O\left(E^3\right) \tag{9}$$

Since the electric field strength E (r,ω) and displacement field D (r,ω) are related to the scalar dielectric constant of the microscopic and isotropic material, ε (r,ω)−χ and the higher order term can be neglected. In low loss dielectric materials, ε(r) can be treated as purely real, thus producing the electric field density written as

$$\mathrm{D(r) = \varepsilon\ (r)\ E\ (r)} \tag{10}$$

In addition, for most dielectric material, the magnetic permeability, μ_r, is approximately equal to 1, giving the magnetic flux density, B, equal to the magnetic field strength, H. The flux density of the dielectric material can be written as D = ε.E where the permittivity, ε, is real. Therefore the Maxwell equation can be rewritten as already illustrated in [21, 47] as

$$\nabla \cdot H(r,t) = 0 \tag{11}$$

$$\nabla \cdot \varepsilon(r) E(r,t) = 0 \tag{12}$$

$$\nabla \times E(r,t) + \frac{1}{c}\left(\frac{\partial H(r,t)}{\partial t}\right) = 0 \tag{13}$$

$$\nabla \times H(r,t) - \frac{\varepsilon(r)}{c}\left(\frac{\partial E(r,t)}{\partial t}\right) = 0 \tag{14}$$

Then the harmonic mode of the E and H field components propagating in the dielectric medium is considered as

$$\mathrm{H(r,t) = H(r)e^{i\omega t} = 0} \tag{15}$$

and

$$\mathrm{E(r,t)} = \mathrm{E(r)e}^{i\omega t} = 0 \tag{16}$$

By substituting Eqs. (15) and (16) into the Maxwell equations (11)–(14), the equation is deduced to a simple condition (two divergence) as shown below:

$$\nabla \cdot H(r) = \nabla \cdot D(r) = 0 \tag{17}$$

where H(r) and E(r) and the field components at t = 0. By deriving Eqs. (15) and (16) and substituting them into Eqs. (13) and (14), the Maxwell equation will become

$$\nabla \times \left(\frac{1}{\varepsilon(r)} \nabla \times H(r)\right) = \left(\frac{\omega}{c}\right)^2 H(r) \tag{18}$$

Thus Eq. (18) derived has H components which become a master equation for dielectric medium, in particular photonic crystal with only magnetic field, H(r) component. This can also be used to recover an electric field component, E(r), of the Maxwell equation given by

$$E(r) = \left[\frac{-ic}{\omega\varepsilon(r)}\right] \nabla x H(r) \tag{19}$$

The final equation given above (18) and (19) is only used primarily to understand the basic concepts of photonic crystal (PhC) structures. These concepts can also be used for more complex structures such as 2D and 3D PhCs.

4. PhC/PhW micro-cavities

One-dimensional (1D)-PhC micro-cavities embedded in narrow photonic wire have been widely studied. A small shift in the periodic mirrors—in particular one situated in the middle of the periodic mirror—will produce a sharp resonance peak in the middle of stopband. This resonant condition oscillates naturally at certain frequencies with greater amplitudes than others within the system. The Q-factor is particularly useful in determining the qualitative behaviour of a system. For some telecom applications, such as dense wavelength division multiplexing (DWDM), the performance of those resonances is determined by their quality factors and optical transmission at a certain resonance frequency. The quality factor of a system is a dimensionless parameter that defines the first-order behaviour, for the decay, of an oscillating frequency within a micro-cavity. It is characterised by the ratio of the resonant frequency to the bandwidth of the resonance or by the decrease in the amplitude of the wave propagating through a system, within an oscillation period. Equivalently, it compares the frequency at which the system oscillates to the rate of energy dissipated by the system. A higher Q-factor value indicates a lower rate of energy dissipation relative to the oscillation frequency, so the oscillations die out more slowly. For example, a pendulum suspended from a high-quality bearing, oscillating in air, would have a high Q, while a pendulum immersed in oil would have a low one. In optics, the Q-factor is generally given by [49, 50]:

$$Q = \frac{2\pi f_0 E}{P} \tag{20}$$

where ε is the stored energy in the cavity and P is the power dissipated within the cavity, given by

$$P = -\frac{dE}{dt} \tag{21}$$

The Q-factor is equal to the ratio of the resonant frequency to the bandwidth of the cavity resonance shown in **Figure 2**.

Ideally, the average lifetime of a resonant photon in the cavity is proportional to the cavity's *Q*. Resonant systems respond to frequencies close to their natural frequency much more strongly than they respond to other frequencies. A system with a high *Q* resonates with greater amplitude (at the resonant frequency) than one with a low *Q*-factor, and its response falls off more rapidly as the frequency moves away from resonance. Therefore the physical interpretation of resonance is given by its general equation:

$$Q = \frac{f_0}{\Delta f} \tag{22}$$

where f_0 is the central frequency of the resonance and Δf is the frequency difference within at 3 dB points or ½ of the total energy stored in the micro-cavity system. In this present work, several different types of resonator have been studied, namely, waveguide Bragg gratings and 1D PhC/PhW waveguides—as shown in **Figure 3**. Unlike the Bragg grating waveguide [51], which has a rectangular recess embedded on a photonic wire waveguide, a single row of holes is used as a set of mirrors.

This structure consists of a single row of holes drilled in the 500 nm width of wire waveguides. Those holes acted as a periodic mirror where light impinging on the PhC bounced back provide a band gap where light is forbidden to propagate at certain frequency. A spacer was introduced symmetrically between the periodic mirrors—thus producing a narrow resonance in the transmission. The concepts for this kind of structure were proposed by Krauss and Foresi [5, 20]. But the Q-factor at this [43] resonance condition obtained was small (~500). The PhC hole mirrors resulted in a wide stopband (approximately 182 nm), using eight PhC mirrors holes, whereas 32 period waveguide Bragg gratings were used and showed a narrower stopband of approximately 88 nm. This difference is due to the fact that the light was coupled more strongly in the periodic hole mirrors—where 95% of the light was reflected with the periodic hole arrangement, as compared to the waveguide Bragg

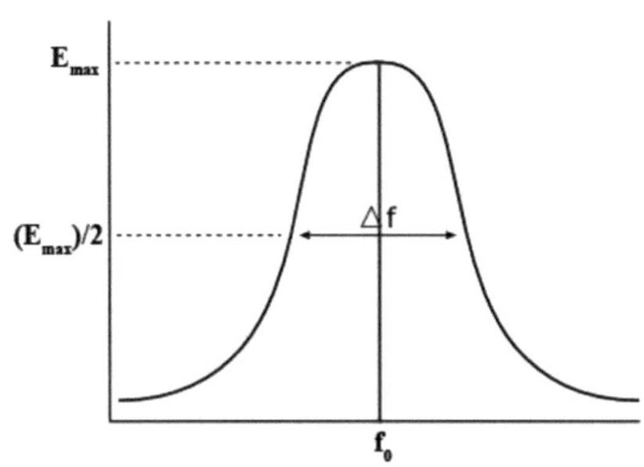

Figure 2.
A typical resonance frequency resulted from micro-cavity structures defined by the central resonance frequency, f_0, and the bandwidth of the frequency at 3 dB points (energy at the steady state condition).

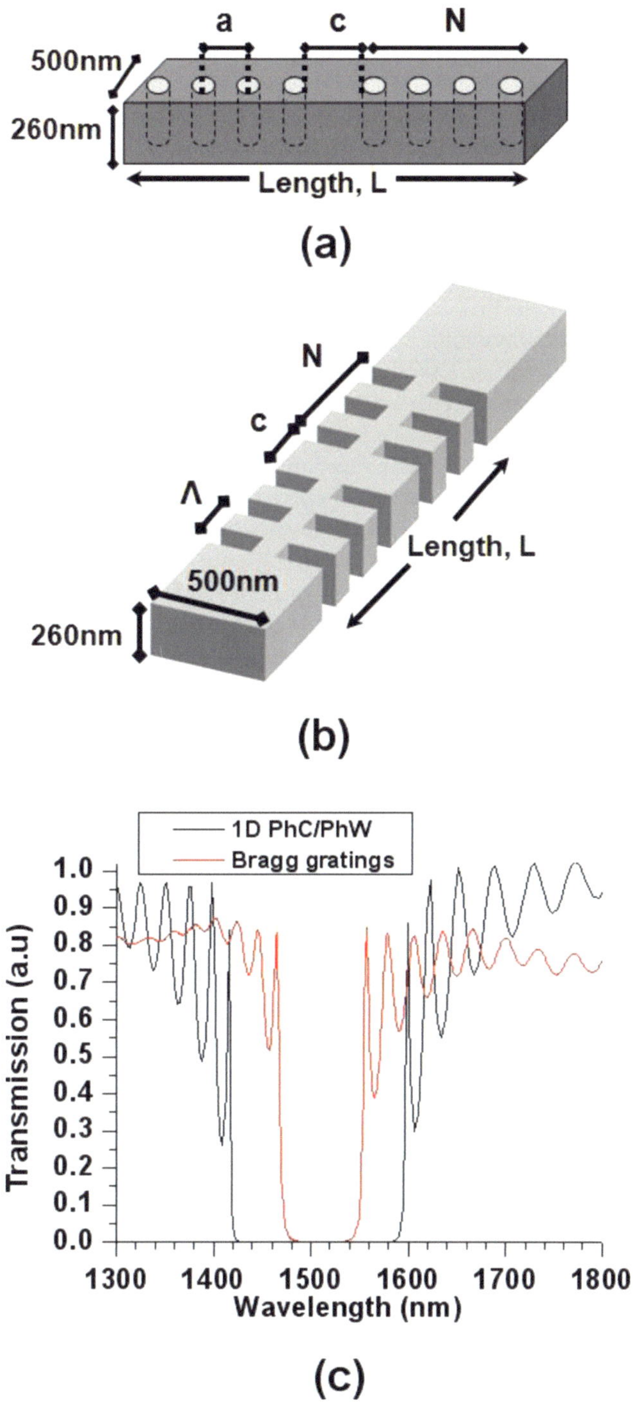

Figure 3.
Different types of PhW waveguides micro-cavity. (a) 1D PhC/PhW waveguides with cavity length (distance between two hole edges of the hole spacer) c, hole periodic spacing (distance between centre-to-centre hole), a and number of periodic holes, N (b) PhW Bragg gratings waveguides with cavity length, c, period, Λ, and number of recess period, N, and (c) transmission spectra of Bragg grating waveguides and 1D PhC/PhW.

gratings where ~82% light was reflected back. Hole gratings show more pronounce stopbands compared to their counterpart. As mentioned before, stronger reflection was observed for the hole grating. The hole gratings have a bigger stopband of approximately 180 nm, which is useful for some filter designs and some optical communications applications. This wide stopband may be compared with the limited bandwidth of the stopband or may be contrasted with the significantly smaller stopband of the rectangular recess grating.

In addition, for this grating condition, the total length, L, of the waveguide Bragg grating of ~11 μm is longer by a factor of four in order to achieve a practical stopband spectrum, as compared with the hole grating structures (~3 μm). The present work will demonstrate the design, fabrication and characterisation of the 1D PhC-based micro-cavity, which is potentially useful for wavelength division multiplexing (WDM) in PhC devices. A single row of PhC holes is embedded in a narrow photonic wire waveguide to allow sufficient optical coupling for integration with other photonic devices. This thesis will address the importance of using a combination of hole tapering with a different hole diameter at the interface between the un-patterned wire and the cavity mirror, as well into the micro-cavity region—in order to achieve large optical transmission together with a high-resonance Q-factor value. Achieving high Q-factor together with large optical transmission remains a significant challenge. The key points towards designing an ultrahigh Q-factor device that confines light in such a small volume lie in reducing the modal mismatch between the un-patterned wire and the PhC or grating sections. Therefore, designing a tapered structure to reduce the modal mismatch at the interfaces between the mirror region and the PhW waveguide sections is necessary. One of the approaches used to overcoming this situations is the use of a taper structure consisting of holes with different sizes through progressive increase of the hole size into the mirror region [52]. On the other hand, the same model has also been used with short taper sections incorporated into a 1D micro-cavity-based system [53]. Using these concepts, the impact of progressive tapering using different hole diameters has shown a huge improvement in enhancing the quality factor of the micro-cavity [54, 55].

Therefore, 1D PhC/PhW micro-cavities can provide higher optical confinement in smaller volumes that are closer to the theoretical value of 0.055 $(\lambda/n)^3$ [54]—which has a great potential in high-index contrast materials such as silicon-on-insulator (SOI) to be used in some telecommunication applications such as DWDM, add-drop filter switching experiments, slow light and non-linear optics.

5. The finite-difference time-domain (FDTD) approach

There have been several methods used for computational purposes, especially for modelling photonic crystal structures and photonic wire waveguides. The finite-difference time-domain (FDTD) approach is a commonly used technique because it provides both the spatial and temporal properties of the structure with a single calculation, making it suitable for the analysis of many structures. However it requires a lot of time to compute a single run. This technique uses the famous Maxwell's equations based on the Yee mesh [56], published in 1966. Yee has proposed this technique in order to derive a numerical scattering problem and electromagnetic absorption on the basis of Maxwell's equations. The computational domain was first established, in order to determine the physical region within which the calculation will be performed. The electric field, E, and the magnetic field, H, are distinguished at every point within the domain by specifying the material used at each domain point (in *xyz* directions). The materials involved could be free space (air), metal or dielectric material. A light source in the form of a plane wave is then impinging on the chosen material. Later in 1994, the technique called the perfectly matched layer (PML) boundary condition was introduced [57]. It was used as an absorption mechanism for electromagnetic wave incident on the edge of the computational domain in space. The FDTD method can be implemented in either 2D or 3D computations—but it requires a lot of memory and power consumption for a single computational run, especially for a large device in 3-D. 2D FDTD reduces time and memory requirement significantly. It employs a refractive index approximation or average refractive index of the slab—called effective index method (EIM). By using this method, the cross-sectional index profile is usually transformed to the one-dimensional index profile by using EIM [58, 59]. In the EIM approach, the eigenvalue of the equivalent slab waveguide is an approximate index value of the original waveguide. Although the EIM approach provides a good approximation, it still suffers from errors in the vicinity of the cut-off [60–63]. At the beginning of this present work, this method is used to investigate the preliminary behaviour of the device with the assumption that losses are negligible. In order to reduce simulation time and power consumption, 2D FDTD approach was initially used throughout the course to analyse the general optical behaviour of the device structures—implementing EIM. Since EIM method is only an approximation of the actual refractive index obtained by taking into account the whole ridge waveguide structures, at least a small discrepancy between the simulations measured results is very much to be expected. On the other hand, the 3D FDTD method can give a better estimate of the properties, although it is time- and power-consuming, which is still a major concern.

During this present work, different types of commercial software have been used. The Fullwave RSoft computational software has been used at the beginning of this work, where only 2D computation was deeply explored due to the longer time and high power consumption for 3D FDTD. Based on the concept proposed by Yee [56], several key pieces of information are needed to solve the basic propagation problem in optical waveguide which comprised of:

- The refractive index distribution, n(x,y,z)
- Electromagnetic field excitation (plane wave or Gaussian)
- Finite computational domain in x, y and z direction
- The boundary of PML layer
- Spatial grid size, Δx and Δy
- Time step, Δt, and the total length of the simulation time

For 2D FDTD computation, the average refractive index, n, or effective index, n_{eff}, of the slab waveguide of a material is used rather than the actual refractive of that particular material. This can be obtained using mode-matching method available in the Fimm-wave® commercial software by Photon Design®. This method includes the approximation of refractive index in both propagation direction of vertical and horizontal confinement of the slab waveguide. The transverse section of the device is first simulated using Fimm-wave® simulation tools. It shows the intensity of light in guiding mode, confinement of light inside the slab and the effective index, n_{eff}. It also shows the leaky region where light is not confined inside the slab. **Figure 4** shows the contour plot of the TE fundamental mode of the waveguide. It shows the intensity of light confinement along the core at 1.52 μm wavelength at different etching depths. It is suggested that the different etching depths will give rise to the abrupt change of the effective index, n_{eff}, at the

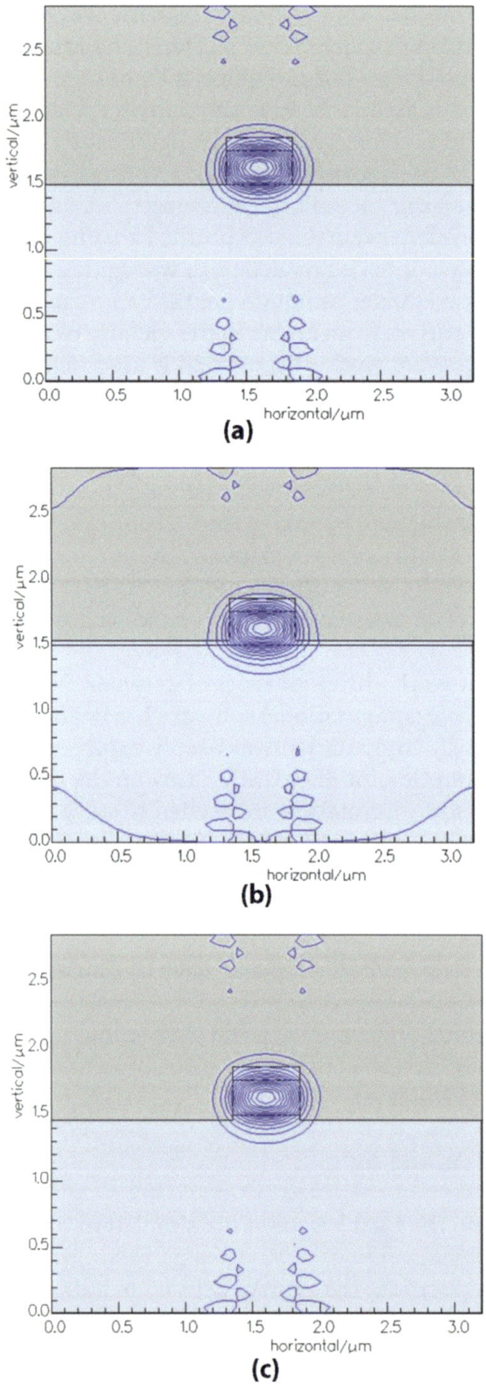

Figure 4.
Contour plot of the TE fundamental mode intensity. (a) Fully etched. (b) Shallow etched. (c) Deep etched.

boundary of silicon core and silica cladding (lower cladding) where some of the light are reflected back into the cladding (backscattering). This can be improved by etching slightly deeper into the lower cladding by around 20–40 nm, thus reducing scattering losses. The effective index calculated using the Fimm-wave™ simulation tool for 500 nm wide ridge waveguides at different etching depths is given in **Figure 4(a)** and **(b)**. More profound field intensity is obtained for fully etched silicon where symmetric field distribution is obtained (see **Figure 4(a)**) as compared to shallow- and deep-etched silicon.

Depositing silica on top of the photonic wire can also improve the confinement of TE fundamental mode of the photonic wire significantly. From **Figure 5**, 100, 200 and 400 nm SiO_2 have been deposited on the photonic wire. But to reduce the device preparation complexity and process development, the slab waveguide design based on fully etched silicon is considered throughout this work. The calculated n_{eff} based on this design is 2.97, which will be used for 2D FDTD computation.

The value of n_{eff} is fed into the full-wave simulation tool by using either pulsed or continuous Gaussian source for slab waveguide. The finite computational domain is optimised in space covering the area between 10 and 20 μm in length and 2 μm in

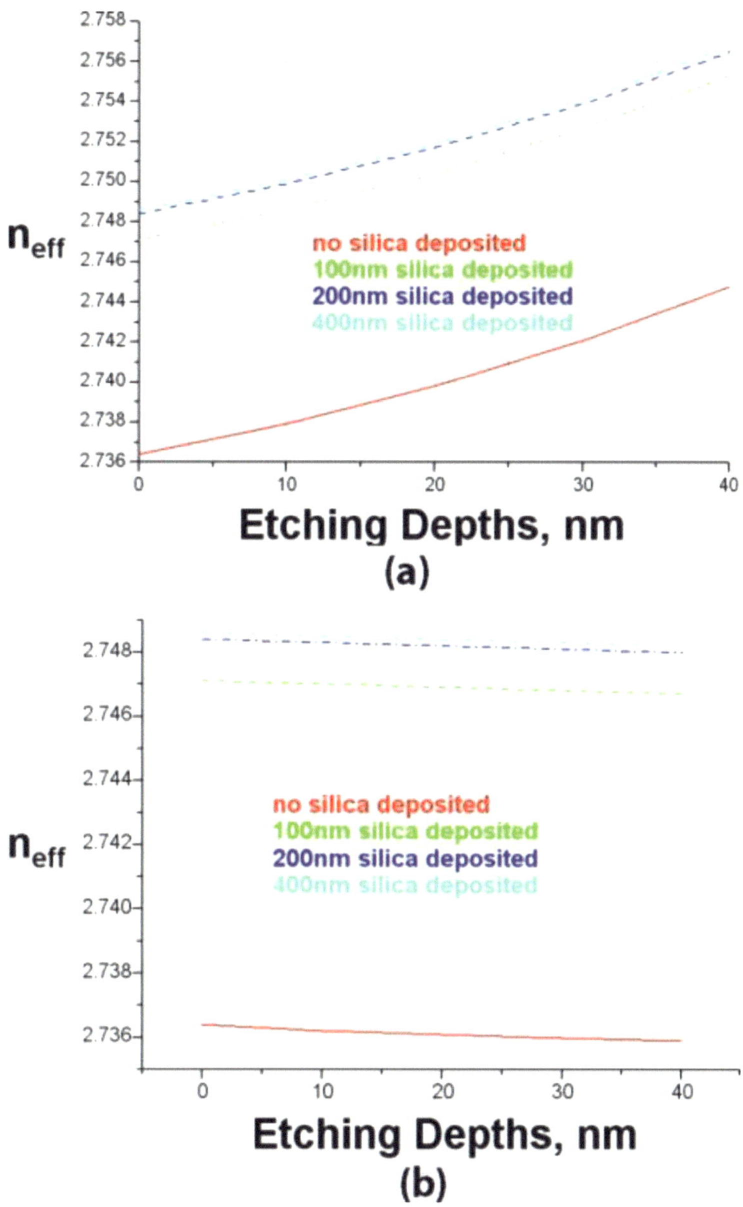

Figure 5.
The effective index, n_{eff}, at different etching depths for symmetric (silica deposition on top) and asymmetric waveguide (no silica deposition). (a) Shallow etched. (b) Deep etched.

width since large computational area will contribute to a longer simulation time and also consume more memory and power.

The space must be exactly proportionate to the size of the optical waveguide. The thickness of PML required for the device operating at around 1.52 μm wavelength is 0.5 μm to provide better electromagnetic wave absorption at the boundary. Other parameters that contribute to the accuracy of the simulation are determined by the choice of the spatial grid or mesh size, where smaller grid spacing gave more accurate computation. In other words, the closer the resolution in simulation to the actual device, the more accurate simulation will be established.

During the second half of the computational process, the author has used other commercially available software to compute all the device structures. This software is found to be more accurate than the previous simulation tool. On the other hand, by using the same parameters in the crystal wave tool as previously used in full-wave, the simulation time has been reduced by a factor of five, and the result obtained is closer to the measured result as shown in one of the example in **Figure 6**. The comparison is made by using 12-period 1D PhC mirrors with diameters of 350 nm and periodic spacing of 360 nm. By looking at the band edge location of the measured result in **Figure 6**, the 2D FDTD crystal wave shows closer result (band edge) than the one computed using RSoft tools where the deviation of 83 nm is observed between the simulation and the measured result. But this is understood to be due to small deviation in the dimension of the structures produced after fabrication process as real devices.

No further investigation is made in reference to the discrepancy between the different examples of commercial software, but the problem has been addressed to the relevant personnel. As a result, based on further tests carried out using different measurements run to compare the results with the simulation, the CrystalWave software have been chosen as the relevant tools that are well-suited to the design structures used throughout this present work.

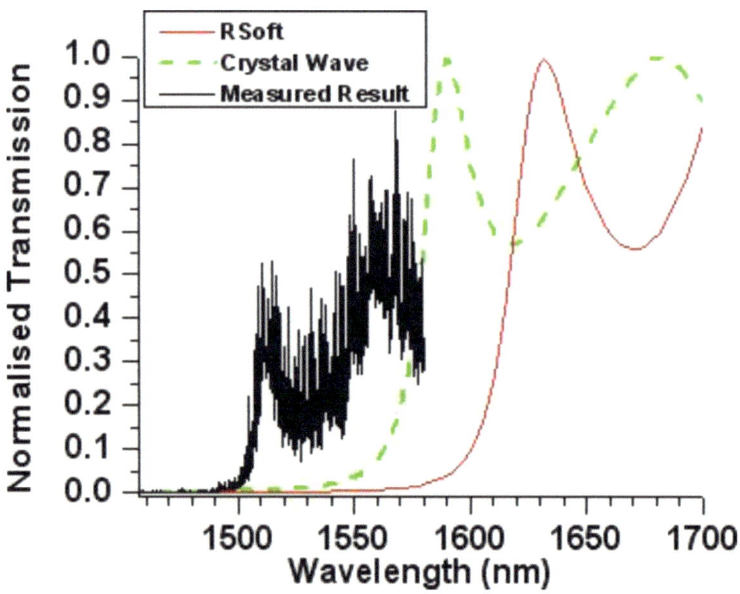

Figure 6.
An example showing a comparison between 2D FDTD computed using different simulations tools (RSoft and crystal wave) with the measured result.

6. Conclusion

In conclusion, the 2D simulation approach is used to produce preliminary designs for the device, together with the employment of an effective index approximation based on the waveguide properties of the base material structure discussed earlier. 2D simulation helps with obtaining a better understanding of the general behaviour of the device—but 3D simulation gives more accurate prediction of the results, at the expense of much greater time and energy consumption. Both 2D and 3D simulations were carried out using the commercial software using finite-difference time-domain (FDTD) approach. This chapter in particular demonstrated the detailed theoretical models of single-row PhC cavities embedded in narrow (typically 500 nm wide) photonic wire waveguides based on silicon-on-insulator (SOI). The device structures have been designed to operate in TE polarisation at wavelengths around 1550 nm. The compactness together with high reflectivity and possibilities for an active tuning capability make the device suitable as a basic building block for incorporation into integrated circuits where several functions are realised on a single chip—i.e. what are commonly known as high-density photonic integrated circuits (PICs). On the other hand, it may also be useful in providing one of the solutions for the design of compact filters for either coarse or dense wavelength division multiplexing situations, for high-speed switching and non-linear optics. For instance, FDTD approach used in this chapter has shown a significantly good agreement with the measured result—thus it can be used as a method to obtain a preliminary result before the actual design for fabrication is proposed.

Author details

Ahmad Rifqi Md Zain[1,2]* and Richard M. De La Rue[3]

1 Institute of Microengineering and Nanoelectronics, University Kebangsaan Malaysia (UKM), Bangi, Selangor, Malaysia

2 John Mc Kay Lab, School of Engineering and Applied Science (SEAS), Harvard University, Cambridge, MA, United States

3 Schoolf of Engineering, University of Glasgow, Glasgow, United Kingdom

*Address all correspondence to: rifqi@ukm.edu.my

References

[1] Miller SE. Integrated optics: An introduction. Bell System Technical Journal. 1969;**48**:2059-2070

[2] Yariv A, Leite RCC. Dielectric-waveguide mode of light propagation in p-n junctions. Applied Physics Letters. 1963;**2**:55

[3] Bond WL, Cohen BG, Leite RCC, Yariv A. Observation of the dielectric - waveguide mode of light propagation in p-n junctions. Applied Physics Letters. 1963;**2**:57

[4] Nelson DF, Reinhardt FK. Light modulation by the electro optic effect in reverse-biased GaP p-n junction. Applied Physics Letters. 1964;**5**:148

[5] Nathan MI, Dumke WD, Burns G, Dill FH Jr, Lasher G. Stimulated emission of radiation from GaAs p-n junction. Applied Physics Letters. 1962; **1**:62

[6] Basov NG, Vul B, Popov YuM. Zhurnal Eksperimentalnoi i Teoreticheskoi Fizik. 1959;**37**;587-588. [Sov. Phys.—JETP. 1960;**10**:416]

[7] Maiman TH. Stimulated optical radiation by ruby lasers. Nature. 1960; **187**:493

[8] Hall RN, Fenner GE, Kingsley JD, Soltys TJ, Carlson RO. Coherent light emission from GaAs junction. Physical Review Letters. 1962;**9**:366-368

[9] Kogelnik H. An introduction to integrated optics. IEEE Transactions on Microwave Theory and Techniques. 1975;**23**(1):2-16

[10] Adams MJ, Steventon AG, Devlin WJ, Henning ID. Semiconductor lasers for long-wavelength optical fibre communications systems. IEE Materials and Devices Series 4. US: Peter Peregrinus; 1987.

[11] Quist TM, Rediker RH, Keyes RJ, Krag WE, Lax B, McWhorter AL, et al. Semiconductor maser of GaAs. Applied Physics Letters. 1962;**1**:91

[12] Purcell EM, Torrey HC, Pound RV. Resonance absorption by nuclear magnetic moments in a solid. Physical Review. 1946;**69**:37

[13] Purcell EM. Spontaneous emission probabilities at radio frequencies. Physical Review Letters. 1946;**69**:681

[14] John S. Strong localization of photons in certain disordered dielectric superlattices. Physical Review Letters. 1987;**58**(23):2486-2489

[15] Yablonovitch E. Inhibited spontaneous emission in solid-state physics and electronics. Physical Review Letters. 1987;**58**(20):2059-2062

[16] Soref RA, Lorenzo JP. All-silicon active and passive guided-wave components for $\lambda = 1.3$ μm and 1.6 μm. IEEE Journal of Quantum Electronics. 1986;**22**(6):873-839

[17] Jalali B, Yegnanarayanan S, Yoon T, Yoshimoto T, Rendina I, Coppinger F. Advances in silicon-on-insulator optoelectronics. IEEE Journal of Selected Topics in Quantum Electronics. 1998;**4**(6):938-947

[18] Masini G, Colace L, Assanto G. Si based optoelectronics for communications. Material Science and Engineering B. 2002;**89**:2-9

[19] Krauss TF, De la Rue RM. Photonic crystals in the optical regime—Past, present and future. Progress in Quantum Electronics. 1999;**23**:2

[20] Krauss TF. Planar photonic crystal waveguide devices for integrated optics.

Physica Status Solidi A—Applied Research. 2003;**197**(3):688-702

[21] Joannopoulos JD, Meade RD, Winn JN. Photonic Crystals Molding the Flow of Light. Princeton, US: Princeton University Press; 1995

[22] Scherer A, Painter O, Vuckovic J, Loncar M, Yoshie T. Photonic crystals for confining, guiding, and emitting light. IEEE Transactions on Nanotechnology. 2002;**1**(1):4-11

[23] Yoshie T, Vuckovic J, Scherer A, Chen H, Deppe D. High quality two-dimensional photonic crystal slab cavities. Applied Physics Letters. 2001; **79**(26):4289-4291

[24] Park H-G, Hwang J-K, Huh J, Ryu H-Y, Lee Y-H, Kim J-S. Nondegenerate monopole-mode two-dimensional photonic band gap laser. Applied Physics Letters. 2001;**79**(19):3032-3034

[25] Painter O, Vuckovic J, Scherer A. Defect modes of a two-dimensional photonic crystal in an optically thin dielectric slab. Journal of the Optical Society of America B. 1999;**16**(2): 275-285

[26] Ryu H-Y, Notomi M, Lee Y-H. High-quality-factor and small-mode-volume hexapole modes in photonic-crystal-slab nanocavities. Applied Physics Letters. 2003;**83**(21):4294-4296

[27] Jin C, Fan S, Han S, Zhang D. Reflectionless multichannel wavelength demultiplexer in a transmission resonator configuration. IEEE Journal of Quantum Electronics. 2003;**39**(1): 160-165

[28] Chong H, De La Rue RM. Planar photonic crystal microcavities for add/drop filter functionality. In: Post-deadline Paper for 11th European Conference on Integrated Optics (ECIO 2003); Prague, Czech Republic. 2003

[29] Lin SY, Chow E, Johnson SG, Joannopoulos JD. Direct measurement of the quality factor in a two-dimensional photonic-crystal microcavity. Optics Letters. 2001; **26**(23):1903-1905

[30] Loncar M, Nedeljkovic D, Doll T, Vuckovic J, Scherer A, Pearsall T. Waveguiding in planar photonic crystals. Applied Physics Letters. 2000; **77**(13):1937-1939

[31] Olivier S, Benisty H, Smith CJM, Rattier M, Weisbuch C, Krauss TF. Transmission properties of two-dimensional photonic crystal channel waveguides. Optical and Quantum Electronics. 2002;**34**:171-181

[32] Notomi M, Shinya A, Yamada K, Takahashi J, Takahashi C, Yokohama I. Singlemode transmission within photonic bandgap of width-varied single-line-defect photonic crystal waveguides on SOI substrates. Electronics Letters. 2001;**37**(5):293-295

[33] O'Faolain L et al. Low loss propagation in photonic crystal waveguide. Electronic Letters. 2006; **25**:42

[34] Armani DK, Kippenberg TJ, Spillane SM, Vahala KJ. Ultralow-threshold microcavity Raman laser on a microelectronic chip. Nature. 2003;**421**: 925

[35] Bogaerts W, Taillaert D, Luyssaert B, Dumon P, Van Campenhout J, Bienstman P, et al. Basic structures for photonic integrated circuits in silicon-on-insulator. Optics Express. 2004;**12**(8): 1583-1591

[36] Ahmad RU, Pizzuto F, Camarda GS, Espinola RL, Rao H, Osgood RM. Ultracompact corner-mirrors and T-branches in silicon-on-insulator. IEEE Photonics Technology Letters. January, 2002;**14**(1):65-67

[37] Ohno F, Fukuzawa T, Baba T. Mach-Zehnder interferometers composed of μ-bends and μ-branches in a Si photonic wire waveguide. Japanese Journal of Applied Physics. 2005;**44** (7A):5322-5323

[38] De La Rue R, Chong H, Gnan M, Johnson N, Ntakis I, Pottier P, et al. Photonic crystal and photonic wire nano-photonics based on silicon-on-insulator. New Journal of Physics. 2006; **8**:256

[39] Camargo EA, Chong HMH, De La Rue RM. Highly compact asymmetric Mach-Zehnder device based on channel guides in a two-dimensional photonic crystal. Applied Optics. 2006;**45**: 6507-6510

[40] Camargo EA, Chong HMH, De La Rue RM. Four-port coupled channel-guide device based on 2D photonic crystal structure. Photonics and Nanostructures—Fundamentals and Applications. 2004;**2**(3):207-213

[41] Zhang H, Gnan M, Johnson NP, De La Rue RM. Ultra-small Mach-Zehnder interferometer devices in thin silicon-on-insulator. In: Integrated Photonics Research and Applications (IPRA); Uncasville, Conn., USA. 2006

[42] Zhao CZ, Li GZ, Liu EK, Gao Y, Liu XD. Silicon-on-insulator Mach-Zehnder wave-guide interferometers operating at 1.3 μm. Applied Physics Letters. 1995; **67**(17):1735

[43] Foresi JS, Villeneuve PR, Ferrera J, Thoen ER, Steinmeyer G, Fan S, et al. Photonic bandgap microcavities in optical waveguides. Nature. 1997;**390**: 143-145

[44] Sakai A, Hara G, Baba T. Propagation characteristics of ultrahigh-delta optical waveguide on silicon-on-insulator substrate. Japanese Journal of Applied Physics Letters. 2001;**40**(4B): L383-L385

[45] Xia F, Sekaric L, Vlasov YA. Ultracompact optical buffers on a silicon chip. Nature Photonics. 2007;**1**(1):65-71

[46] Gnan M, Thoms S, Macintyre DS, De La Rue RM, Sorel M. Fabrication of low-loss photonic wires in silicon-on-insulator using hydrogen silsesquioxane electron-beam resist. Electronics Letters. 2008;**44**:115-116

[47] Jackson JD. Classical Electrodynamics. New York: John Wiley; 1962

[48] Jackson JD. Classical Electrodynamics. 3rd ed. New York: Wiley; 1998. p. 177

[49] Jackson RG. Novel Sensors and Sensing. CRC Press; 2004. p. 28. ISBN 075030989X

[50] Crowell B. Vibrations and Waves. Light and Matter Online Text Series. 2006. Ch. 2

[51] Gnan M, Bellanca G, Chong HMH, Bassi P, De La Rue RM. Modelling of photonic wire Bragg gratings. Optical and Quantum Electronics. 2006;**38** (1–3):133-148

[52] Lalanne P, Talneau A. Modal conversion with artificial materials for photonic-crystal waveguides. Optics Express. 2002;**10**:354-359

[53] Peyrade D, Silberstein E, Lalanne P, Talneau A, Chen Y. Short Bragg mirrors with adiabatic modal conversion. Applied Physics Letters. 2002;**81**: 829-831

[54] Velha P, Rodier JC, Lalanne P, Hugonin JP, Peyrade D, Picard E, et al. Ultra high reflectivity photonic bandgap mirrors in a ridge SOI waveguide. New Journal of Physics—IOPscience. 2006; **8**(204):1-13

[55] Lalanne P, Hugonin JP. Bloch-wave engineering for high-Q, small-V

microcavities. IEEE Journal of Quantum Electronics. Nov. 2003;**39**(11): 1430-1438

[56] Yee KS. Numerical solution of initial boundary value problems involving Maxwell's equations in isotropic media. IEEE Transactions on Antennas and Propagation. 1966;**14**:302-307

[57] Berenger JP. A perfectly matched layer for the absorption of electromagnetic waves. Journal of Computational Physics. 1994;**114**(2): 185-200

[58] Buus J. The effective index method and its application to semiconductor lasers. IEEE Journal of Quantum Electronics. 1982;**18**(7):1083-1089

[59] Chung Y, Dagli N. An assessment of finite difference beam propagation method. IEEE Journal of Quantum Electronics. Aug. 1990;**26**(8):1335-1339

[60] Cheng YH, Lin WG. Investigation of rectangular dielectric waveguides: An iteratively equivalent index method. IEE Proceedings. 1990;**137**(5):323-329

[61] Furuta H, Noda H, Ihaya A. Novel optical waveguide for integrated optics. Applied Optics. 1974;**13**:323-326

[62] Kumar A, Thyayarajan K, Ghatak AK. Analysis of rectangular core dielectric waveguide: An accurate perturbation approach. Optics Letters. 1983;**8**:63-65

[63] Kim CM, Jung BG, Lee CW. Analysis of dielectric rectangular waveguide by modified effective index method. Electronic Letters. 1986;**22**(6): 296-298